THE CARTOON GUIDE TO

GEOMETRY

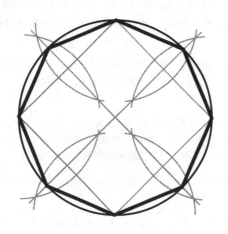

ALSO BY LARRY GONICK

THE CARTOON HISTORY OF THE UNIVERSE, VOLUMES 1–7

THE CARTOON HISTORY OF THE UNIVERSE, VOLUMES 8–13

THE CARTOON HISTORY OF THE UNIVERSE, VOLUMES 14–19

THE CARTOON HISTORY OF THE MODERN WORLD, PART 1

THE CARTOON HISTORY OF THE MODERN WORLD, PART 2

THE CARTOON HISTORY OF THE UNITED STATES

THE CARTOON GUIDE TO ALGEBRA

THE CARTOON GUIDE TO BIOLOGY (WITH DAVE WESSNER)

THE CARTOON GUIDE TO CALCULUS

THE CARTOON GUIDE TO CHEMISTRY (WITH CRAIG CRIDDLE)

THE CARTOON GUIDE TO THE COMPUTER

THE CARTOON GUIDE TO THE ENVIRONMENT (WITH ALICE OUTWATER)

THE CARTOON GUIDE TO GENETICS (WITH MARK WHEELIS)

THE CARTOON GUIDE TO (NON)COMMUNICATION

THE CARTOON GUIDE TO PHYSICS (WITH ART HUFFMAN)

THE CARTOON GUIDE TO SEX (WITH CHRISTINE DEVAULT)

THE CARTOON GUIDE TO STATISTICS (WITH WOOLLCOTT SMITH)

THE ATTACK OF THE SMART PIES

"GONICK'S *THE CARTOON GUIDE TO STATISTICS*...[IS] THE ONLY REFERENCE TEXT FOR MY GENERAL EDUCATION COURSE 'REAL-LIFE STATISTICS: YOUR CHANCE FOR HAPPINESS (OR MISERY).'"
—XIAO-LI MENG, PROFESSOR OF STATISTICS, HARVARD UNIVERSITY

"SO CONSISTENTLY WITTY AND CLEVER THAT THE READER IS BARELY AWARE OF BEING GIVEN A THOROUGH GROUNDING IN THE SUBJECT." —*OMNI* MAGAZINE

"[*THE CARTOON HISTORY OF THE UNIVERSE, BOOK 3, IS*] A MASTERPIECE!" —STEVE MARTIN

"LARRY GONICK SHOULD GET AN OSCAR FOR HUMOR AND A PULITZER FOR HISTORY."
—RICHARD SAUL WURMAN, CREATOR OF THE TED CONFERENCES

GONICK'S CARTOON HISTORIES AND CARTOON GUIDES HAVE BEEN REQUIRED READING IN COURSES AT BISMARCK HIGH SCHOOL, BISMARCK, NORTH DAKOTA; BLOOMSBURG UNIVERSITY; BOSTON COLLEGE; BUCKINGHAM BROWNE & NICHOLS SCHOOL, CAMBRIDGE, MASSACHUSETTS; CALIFORNIA INSTITUTE OF THE ARTS; CALIFORNIA STATE UNIVERSITY AT CHICO; CARNEGIE-MELLON UNIVERSITY; COLUMBIA UNIVERSITY; CORNELL UNIVERSITY; DARTMOUTH COLLEGE; DUKE UNIVERSITY; GIRVAN ACADEMY, SCOTLAND; HARVARD UNIVERSITY; HUMBOLDT STATE UNIVERSITY; HUNTINGDON COLLEGE; ILLINOIS STATE UNIVERSITY; JOHN JAY COLLEGE; JOHNS HOPKINS UNIVERSITY; KENT SCHOOL DISTRICT, KENT, WASHINGTON; KENYON COLLEGE; LANCASTER UNIVERSITY, ENGLAND; LICK-WILMERDING HIGH SCHOOL, SAN FRANCISCO, CALIFORNIA; LIVERPOOL UNIVERSITY, ENGLAND; LOGAN HIGH SCHOOL, LOGAN, UTAH; LONDON SCHOOL OF ECONOMICS; LOUISIANA STATE UNIVERSITY; LOWELL HIGH SCHOOL, SAN FRANCISCO, CALIFORNIA; MARIN ACADEMY; MARQUETTE HIGH SCHOOL, CHESTERFIELD, MISSOURI; MIT; NEW YORK UNIVERSITY; NORTH CAROLINA STATE UNIVERSITY; NORTHWESTERN UNIVERSITY; THE NUEVA SCHOOL, HILLSBOROUGH, CALIFORNIA; THE OHIO STATE UNIVERSITY; PENNSYLVANIA STATE UNIVERSITY; PHILIPPINE HIGH SCHOOL, DILMAN, PHILIPPINES; REDBUD ACADEMY, AMARILLO, TEXAS; ROCHESTER INSTITUTE OF TECHNOLOGY; RUTGERS UNIVERSITY; SAN DIEGO STATE UNIVERSITY; SAN DIEGO SUPERCOMPUTER CENTER; SOUTHEAST MISSOURI STATE UNIVERSITY; SOUTHWOOD HIGH SCHOOL, SHREVEPORT, LOUISIANA; ST. IGNATIUS COLLEGE PREPARATORY, SAN FRANCISCO, CALIFORNIA; STANFORD UNIVERSITY; SWARTHMORE COLLEGE; TEMPLE UNIVERSITY; UNIVERSITEIT UTRECHT, NETHERLANDS; THE UNIVERSITY OF ALABAMA; UNIVERSITY OF CALIFORNIA AT BERKELEY, LOS ANGELES, SANTA BARBARA, SANTA CRUZ, AND SAN DIEGO; THE UNIVERSITY OF CHICAGO; THE UNIVERSITY OF EDINBURGH, SCOTLAND; UNIVERSITY OF FLORIDA; UNIVERSITY OF IDAHO; UNIVERSITY OF ILLINOIS; UNIVERSITY OF LEICESTER, ENGLAND; UNIVERSITY OF MARYLAND; UNIVERSITY OF MIAMI, FLORIDA; THE UNIVERSITIES OF MICHIGAN, MISSOURI, NEBRASKA, NEW BRUNSWICK, SCRANTON, SOUTH FLORIDA, TEXAS, TORONTO, WASHINGTON, AND WISCONSIN; YALE UNIVERSITY; AND MANY MORE INSTITUTIONS OF HIGHER AND LOWER EDUCATION!

THE CARTOON GUIDE TO
GEOMETRY

LARRY GONICK

WILLIAM MORROW
An Imprint of HarperCollins*Publishers*

TO EUCLID, AND ALL MY OTHER MATH TEACHERS

HarperCollins books may be purchased for educational, business, or sales promotional use. For information, please email the Special Markets Department at SPsales@harpercollins.com.

FIRST EDITION

Photo credits: p. 125, from Wikipedia article "Truss"; pp. 157, 161, 211, photos courtesy of the author; p. 192, from Wikipedia article "Ferris Wheel (1893)"; p. 221, photo by Brian Ruppert, from Wikimedia Commons (https://commons.wikimedia.org/wiki/File:Middleton_Community_Orchestra_-_10338954426.jpg), reproduced under the Creative Commons License (https://creativecommons.org/licenses/by/4.0/).

Library of Congress Cataloging-in-Publication Data has been applied for.

ISBN 978-0-06-315757-6

23 24 25 26 27 LBC 5 4 3 2 1

CONTENTS

CHAPTER 1 1
    GEOMETRY AND NATURE

CHAPTER 2 13
    BASIC INGREDIENTS

CHAPTER 3 29
    NUMBERS AND LINES

CHAPTER 4 37
    COMPASS AND CIRCLE

CHAPTER 5 47
    ANGLES

CHAPTER 6 59
    THE TRIANGLE

CHAPTER 7 75
    INEQUALITIES IN TRIANGLES

CHAPTER 8 83
    CLASSICAL CONSTRUCTIONS

CHAPTER 9 91
    THE INTERSECTION PROBLEM

CHAPTER 10 99
    BEING PARALLEL

CHAPTER 11 109
    TRIANGLES IN A FLAT WORLD

CHAPTER 12 117
    ONE MORE SIDE!

CHAPTER 13 131
    AREA

CHAPTER 14 147
    THE PYTHAGOREAN THEOREM

CHAPTER 15 157
    SIMILARITY

CHAPTER 16 175
    SCALING AREAS

CHAPTER 17 183
    CIRCLING BACK TO CIRCLES

CHAPTER 18 195
    THE GEOMETRIC MEAN

CHAPTER 19 205
    THE GOLDEN TRIANGLE

CHAPTER 20 213
    TRIG TRICKS

CHAPTER 21 223
    POLYGONS

SOLUTIONS TO SELECTED PROBLEMS 243

ACKNOWLEDGMENTS 252

INDEX 253

Chapter 1

GEOMETRY AND NATURE

"GEOMETRY" ORIGINALLY MEANT "EARTH-MEASUREMENT," AS IN THE SHAPES, SIZES, AND DESIGNS OF FIELDS, BUILDINGS, BRIDGES, DECORATIONS, AND OTHER THINGS HERE ON EARTH.

DOES THAT MEAN WE CAN'T USE IT?

OKAY, FINE, IT PROBABLY WORKS IN OTHER PLACES, TOO.

GEOMETRY EVEN APPEARS IN KITCHENWARE. SAY A CAKE RECIPE CALLS FOR A 9-INCH SQUARE PAN, BUT MOMO'S PANS ARE ALL ROUND. HOW BIG A CIRCLE SHOULD SHE USE?

GEOMETRY HAS ALWAYS BEEN PART OF CONSTRUCTION. PHARAOH'S ARCHITECTS, FOR EXAMPLE, HAD TO ESTIMATE HOW MUCH STONE WOULD GO INTO A PYRAMID 300 CUBITS ON A SIDE.

MEANWHILE, PHARAOH'S **SURVEYORS** REGULARLY MEASURED THE KINGDOM'S FARMLAND FOR TAX PURPOSES. HOW DID THEY FIND THE AREA OF A FOUR-SIDED FIELD?

IF THE FOUR SIDES IN ORDER ARE A, B, a, b, THE ANCIENT EGYPTIAN TAX BUREAU USED THIS FORMULA:

$$\text{Area} = \frac{(A+a)}{2} \cdot \frac{(B+b)}{2}$$

THAT IS, FIRST AVERAGE THE OPPOSITE SIDES, THEN MULTIPLY THE RESULTS.

AT AROUND THE SAME TIME, IN PRESENT-DAY IRAQ, BABYLONIAN GEOMETERS INVENTED THE IDEA OF DIVIDING A CIRCLE INTO 360 EQUAL PARTS, OR **DEGREES.**

IT WAS A BRILLIANT CHOICE OF NUMBERS, BECAUSE 360 CAN BE DIVIDED IN SO MANY WAYS. SEVERAL IMPORTANT "PIE SLICES" MEASURE A WHOLE NUMBER OF DEGREES.

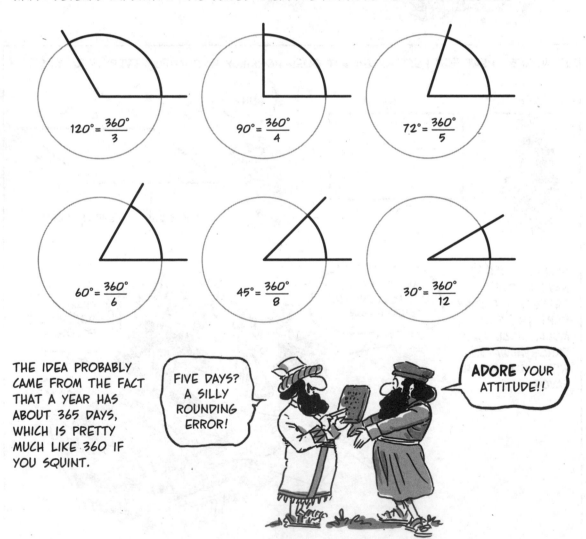

$$120° = \frac{360°}{3}$$

$$90° = \frac{360°}{4}$$

$$72° = \frac{360°}{5}$$

$$60° = \frac{360°}{6}$$

$$45° = \frac{360°}{8}$$

$$30° = \frac{360°}{12}$$

THE IDEA PROBABLY CAME FROM THE FACT THAT A YEAR HAS ABOUT 365 DAYS, WHICH IS PRETTY MUCH LIKE 360 IF YOU SQUINT.

THE BABYLONIANS ALSO SOLVED—APPROXIMATELY—THIS PROBLEM: GIVEN A SQUARE OF A GIVEN SIDE, HOW LONG IS ITS **DIAGONAL,** THE DISTANCE BETWEEN OPPOSITE CORNERS? (A SQUARE HAS FOUR EQUAL SIDES AND FOUR 90-DEGREE CORNER ANGLES.)

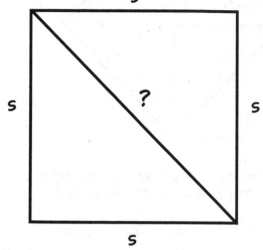

MULTIPLY THE SIDE BY **1.4142!**

OOH! MYSTERIOUS!

A SURVIVING TEXT SAYS THAT IF **s** IS 30 UNITS LONG, THEN THE DIAGONAL MEASURES **42.426** (= 30 × 1.4142) UNITS.

YOU CAN CHECK THIS YOURSELF. DRAW A LARGE SQUARE ON GRAPH PAPER; MEASURE THE SIDE AND DIAGONAL; DIVIDE THE DIAGONAL BY THE SIDE. YOUR ANSWER SHOULD BE CLOSE TO 1.4142.

YES INDEED! CLOSE ENOUGH!

OKAY! FINE! WONDERFUL! BUT **WHERE DID THAT NUMBER 1.4142 COME FROM?**

THE TRICK IS TO NOTICE THAT d, THE DIAGONAL, IS THE SIDE OF A SQUARE WITH **TWICE THE AREA** OF THE ORIGINAL SQUARE.

THE DIAGONAL CUTS THE SMALL SQUARE INTO TWO IDENTICAL TRIANGLES; THE LARGER SQUARE CONTAINS FOUR COPIES OF THIS TRIANGLE, SO THE LARGE SQUARE HAS DOUBLE THE AREA OF THE SMALL ONE.

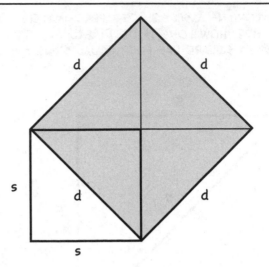

NOW, THE AREA OF A SQUARE, I HOPE YOU REMEMBER, IS THE PRODUCT OF THE SIDE TIMES ITSELF. THE SMALL SQUARE HAS AREA s^2. THE LARGE SQUARE HAS AREA d^2.

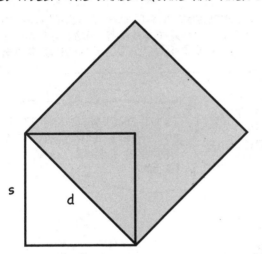

AND ONE IS TWICE THE OTHER.

$$d^2 = 2s^2$$

TAKING THE SQUARE ROOT OF EACH SIDE,

$$d = \sqrt{2s^2}$$

$$d = s\sqrt{2}$$

THE DIAGONAL IS THE SIDE **TIMES THE SQUARE ROOT OF 2.**

THE SQUARE ROOT OF 2 IS NOT EXACTLY EQUAL TO THE BABYLONIANS' 1.4142, BUT IT'S AWFULLY CLOSE.

$$1.4142 \times 1.4142$$
$$=$$
1.99996164

YEAH, GOOD ENOUGH!

ISN'T ANYONE IN THIS $%#$ CIVILIZATION A **PERFECTIONIST?**

NOW THE STORY OF EARTH-MEASUREMENT TAKES A WEIRD TWIST, IN THE HANDS OF A MATHEMATICAL CULT LEADER NAMED **PYTHAGORAS OF SAMOS** (c. 570 BCE–c. 490 BCE). PYTHAGORAS WAS HARD-CORE: HE BELIEVED THAT THE ONLY NUMBERS WERE **WHOLE NUMBERS**.

I'M 100% RATIONAL, AND SO IS EVERYTHING ELSE!

PYTHAGORAS TOOK IT ON FAITH THAT THE **WHOLE WORLD** COULD BE DESCRIBED BY **WHOLE NUMBERS** (INCLUDING FRACTIONS, OR RATIOS OF WHOLE NUMBERS).

AND YOU BELIEVE THIS **WHY?**

FOR, AHEM, A NUMBER OF REASONS.

REASON #1: I'M A CRANK.

SO IT CAME AS A SHOCK WHEN HE OR ONE OF HIS STUDENTS PROVED THAT THE SQUARE ROOT OF TWO IS **IRRATIONAL**. $\sqrt{2}$ **CANNOT BE EXPRESSED AS A FRACTION.*** HERE IS A LENGTH THAT **ISN'T A NUMBER**, FROM THIS GREEK'S POINT OF VIEW.

UNNATURAL! ABSURD! OFFENSIVE! IRRITATING! AND WORST OF ALL— IT'S NOT **WRONG!!**

1

$\sqrt{2}$

1

*YOU CAN FIND A PROOF ONLINE.

THE GREEKS SAID THAT THE SIDE AND DIAGONAL WERE "INCOMMENSURABLE." THE TWO LENGTHS CAN'T BE "MEASURED TOGETHER."

IN-CONVENIENT.

UN-BELIEVABLE.

AND SO PAINFUL...

NO RULER THAT MEASURES THE SIDE OF A SQUARE AS A WHOLE NUMBER OF UNITS CAN ALSO MEASURE THE DIAGONAL AS A WHOLE NUMBER OF UNITS.

WHAT GOOD ARE YOU?

AS FAR AS PYTHAGORAS WAS CONCERNED, THIS THING WE CALL $\sqrt{2}$ IS NO NUMBER. AND YET, THERE IT WAS, A PERFECTLY GOOD LENGTH. APPARENTLY NUMBERS—WHOLE NUMBERS—DON'T DESCRIBE NATURE, AFTER ALL!

GOOD-BYE, OLD FRIEND!

HOW FAR DID THIS MAD IMPULSE CARRY THEM?

WHUMP!

THIS FAR!

AROUND 300 BCE, A GREEK PROFESSOR, **EUCLID** BY NAME, WROTE A TOME THAT SYSTE-MATICALLY DEVELOPED A "PURE" GEOMETRY **WITHOUT NUMBERS.** STARTING FROM A FEW BASIC ASSUMPTIONS ABOUT POINTS AND LINES, EUCLID WENT ON TO DRAW MIND-BOGGLING COMPLICATIONS.

THE ONLY NUMBERS ARE THE CHAPTER HEADINGS!

HIS BOOK, *THE ELEMENTS,* BECAME **THE** GEOMETRY TEXTBOOK FOR THE NEXT TWENTY-TWO CENTURIES, MORE OR LESS.

IT'S A BEST-SELLER!

WHERE'S MY %$#& TEEVEE DEAL?

IN THIS BOOK, WE DON'T FOLLOW EUCLID'S **OUTLINE,** BUT WE DO RESPECT HIS **METHOD:** THE LOGICAL REASONING HE USED AT EVERY STEP... THE PAINSTAKING CARE HE TOOK TO ENSURE THAT EACH NEW RESULT DEPENDED ONLY ON WHAT CAME BEFORE... THE FIRM FOUNDATION HE ESTABLISHED OF... OF... OF...

YEAH, WHAT'S DOWN THERE, ANYWAY?

Questions

1. LET'S TRY DOING THAT TILTED-SQUARE TRICK STARTING WITH A **RECTANGLE** INSTEAD OF A SQUARE. FOR INSTANCE, HOW LONG IS THE DIAGONAL OF A 2 × 5 RECTANGLE?

AS BEFORE, WE MAKE A SQUARE BASED ON THE DIAGONAL. ITS AREA IS c^2.

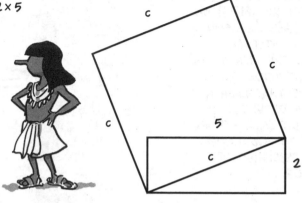

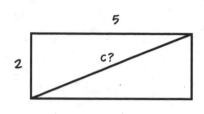

AGAIN WE TRY TO FILL IT WITH FOUR TRIANGLES HALF THE SIZE OF THE ORIGINAL RECTANGLE, BUT NOW THIS LEAVES A SQUARE HOLE IN THE CENTER. ITS SIDE IS $5 - 2 = 3$, SO ITS AREA IS $3^2 = 9$.

RESULT: THE TILTED SQUARE IS THE SUM OF TWO 2 × 5 RECTANGLES AND A 3 × 3 SQUARE. WHAT IS c?

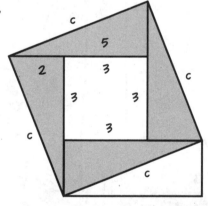

$$c^2 = 2(10) + 3^2 = 29$$
$$c = \sqrt{29}$$

IF WE REPLACE 2 WITH a AND 5 WITH b, THE DIAGRAM TELLS US THAT

$$c^2 = 2ab + (b-a)^2$$
$$= 2ab + a^2 + b^2 - 2ab$$
$$= a^2 + b^2$$

NOW, **THAT** IS A LOVELY FORMULA, BUT THE WAY WE GOT THERE WAS A LITTLE SLOPPY. FOR ONE THING, WE NEVER DEFINED "RECTANGLE" OR SHOWED THAT THOSE SHADED REGIONS REALLY FIT TOGETHER SO PERFECTLY. WELL, IT TURNS OUT THEY DO... AND LATER WE'LL SHOW HOW TO PROVE THIS FORMULA **FOUR MORE TIMES** IN FOUR DIFFERENT WAYS; IT'S THAT IMPORTANT.

2. ARE YOU READY TO DO SOME GEOMETRY?

Chapter 2
BASIC INGREDIENTS

DEFINE YOUR TERMS, OR DON'T

I WELL REMEMBER MY NINTH-GRADE ENGLISH TEACHER, THE TALL, KEEN-EYED, PREMATURELY BALDING SHOWMAN **ZENO JOHNSON**, AND HIS CONSTANT MANTRA. **"DEFINE YOUR TERMS!!"** HE WOULD BELLOW. "DEFINE YOUR TERMS!!"

OR I'LL THROW CHALK AT YOU!

WITH ALL DUE RESPECT, ZENO, WHEREVER YOU ARE, THAT IS NOT ALWAYS A GOOD IDEA IN GEOMETRY.

TO ILLUSTRATE THE, UH, POINT, CONSIDER HOW THE WORD "POINT" MIGHT BE DEFINED.

THERE'S NO WAY AROUND IT: WHEN WE TRY TO DEFINE THE TERMS WE USE TO DEFINE THE TERMS WE USE (GET THAT?), WE EVENTUALLY FIND OURSELVES GOING IN CIRCLES. THERE ARE ONLY SO MANY WORDS IN THE LANGUAGE.

UNLESS, THAT IS, WE LEAVE A FEW WORDS FOREVER **UNDEFINED** AND DEFINE EVERYTHING ELSE IN TERMS OF THESE. IN THAT CASE, GOOD-BYE, CIRCULAR DEFINITIONS!

IN THIS BOOK, THE WORDS "POINT," "LINE," "PLANE," AND "SPACE," ON WHICH EVERYTHING ELSE IS BUILT, HAVE **NO DEFINITION.**

SO... GEOMETRY HAS A FOUN-DATION OF... THIN AIR?

AT LEAST IT HAS A FOUNDATION.

THIS MAY SEEM LIKE A STRANGE WAY TO PROCEED, BUT THEN, GEOMETRY IS A STRANGE SUBJECT. IT TRULY EXISTS ONLY IN OUR MINDS, NOT IN THE MUCK BELOW.

WHATEVER YOU DO, DON'T LOOK DOWN!

SINCE IT'S ALL IN OUR MINDS ANYWAY, LET'S THINK A LITTLE ABOUT HOW WE IMAGINE POINTS, LINES, PLANES, AND SPACE.

15

LINES A LINE IS ONE-DIMENSIONAL, PERFECTLY STRAIGHT, WITH NO WIDTH OR THICKNESS. ITS LENGTH HAS NO END. A LINE IS LIKE A TAUT THREAD, ENDLESS AND THINNER THAN THIN.

POINTS

POINTS HAVE NO LENGTH, WIDTH, OR DEPTH—NONE— YET, SOMEHOW, THEY'RE EVERYWHERE! THEY'RE ZERO-DIMENSIONAL. NOTH-ING IS SMALLER THAN A POINT, PERIOD.

IS THAT A PERIOD OR A FAT POINT?

WHOA! THEY TOOK THE **EARTH** OUT OF EARTH-MEASUREMENT, TOO!

LINES ARE STRINGS OF POINTS, PLANES ARE SHEETS OF LINES, AND SPACE IS A PILE OF PLANES. EVERYTHING IS SMOOTH, STRAIGHT, FLAT, AND PERFECT. IN OTHER WORDS, THIS IS **NOT** THE LUMPY REAL WORLD.

Postulates

HOW CAN WE EXPRESS THESE VAGUE AND HAZY CONCEPTS PRECISELY? EUCLID BRILLIANTLY DID IT BY MAKING SOME VERY SIMPLE ASSUMPTIONS, KNOWN AS **POSTULATES.**

STUFF THAT NO ONE WOULD EVER ARGUE WITH!

YOUR OPTIMISM IS ADORABLE!

Postulate 1. GIVEN ANY TWO POINTS, THERE IS ONE AND ONLY ONE LINE THAT CONTAINS THEM BOTH. WE SAY, "TWO POINTS DETERMINE A LINE."

ALWAYS THIS:

NEVER THIS:

OR THIS:

Postulate 2. GIVEN ANY POINT P, THERE ARE (AT LEAST) TWO OTHER POINTS SO THAT P IS NOT ON "THEIR" LINE.

THIS ENSURES THAT WE'RE NOT CONFINED TO A SINGLE LINE, WHICH WOULD BE BORING.

IT'S **SO** MUCH MORE INTERESTING NOW!

P

Postulate 3. IF A PLANE CONTAINS TWO POINTS, THEN IT CONTAINS THE LINE THEY DETERMINE.

ALWAYS THIS:

NEVER THIS:

OR THIS:

Postulate 4. IF THREE POINTS ARE NOT ON THE SAME LINE, THERE IS ONE AND ONLY ONE PLANE CONTAINING THEM ALL.

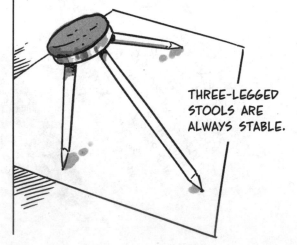

THREE-LEGGED STOOLS ARE ALWAYS STABLE.

HOW CAN SOMETHING BE **TRUE** ABOUT "OBJECTS" THAT DON'T EVEN **EXIST**?

TOUGH QUESTION!

UM... TRUE...

"POSTULATE" IS BOTH A NOUN AND A VERB. AS A VERB, TO POSTULATE SOMETHING MEANS TO ASSUME THAT IT'S TRUE **FOR THE SAKE OF ARGUMENT.**

IN GENERAL, GEOMETRICAL STATEMENTS ARE NOT LIKE REAL-WORLD FACTS; YOU CAN'T VERIFY THEM BY OBSERVATION OR MEASUREMENT. AN EXAMPLE OF A VERIFIABLE STATEMENT IS, "THE EARTH IS ABOUT 25,000 MILES AROUND AT THE EQUATOR." THIS CAN BE MEASURED!

YEP!

A GEOMETRICAL STATEMENT IS MORE LIKE: "IF THE DISTANCE FROM THE CENTER OF A SPHERE TO ITS SURFACE IS 4,000 MILES, THEN THE SPHERE IS ABOUT 25,000 MILES AROUND." THERE'S NO CHECKING THIS—EXCEPT AGAINST THE LOGIC OF GEOMETRY ITSELF. (THE FORMULA ULTIMATELY DEPENDS ONLY ON THE POSTULATES WE ASSUME.)

THERE'S REALITY, AND THERE'S GEOMETRY!*

I'M REALITY, I THINK...

$C = 2\pi r$

*IF SOMETHING GEOMETRICAL TURNED OUT TO BE A BAD FIT WITH REALITY, THOUGH, WE MIGHT WANT TO RETHINK OUR POSTULATES OR OUR IDEAS ABOUT REALITY, OR BOTH.

THE GAME PROCEEDS LIKE THIS: WE PROVE THEOREMS—STATEMENTS ABOUT POINTS, LINES, AND PLANES—USING STEP-BY-STEP REASONING, WITH EACH STEP JUSTIFIED BY THE APPROPRIATE POSTULATE (OR, LATER, BY A PREVIOUSLY PROVED THEOREM).

BEFORE PROVING OUR FIRST THEOREM, HERE'S SOME TERMINOLOGY: IF A AND B ARE TWO POINTS, WE WRITE

$\overline{AB}$

FOR THE (ONE AND ONLY) LINE THEY SHARE. IF C IS ANY OTHER POINT ON $\overline{AB}$, WE SAY THAT A, B, AND C ARE **COLLINEAR**. (THE THREE POINTS SHARE A LINE.) IF C IS NOT ON $\overline{AB}$, THEN A, B, AND C ARE **NONCOLLINEAR**.

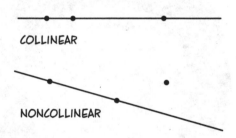

COLLINEAR

NONCOLLINEAR

Theorem 2-1. IF A, B, AND C ARE THREE DISTINCT NONCOLLINEAR POINTS, THEN THE LINES THEY DETERMINE ($\overline{AB}$, $\overline{AC}$, AND $\overline{BC}$) ALL LIE IN THE SAME PLANE.

Proof. WE NUMBER THE STEPS AND JUSTIFY EACH STEP IN THE SECOND COLUMN.

1. A, B, AND C ARE NONCOLLINEAR.   (ASSUMED)

2. A, B, AND C DETERMINE A PLANE P.   (POST. 4)

3. A AND B ARE IN P.   (STEP 2)

4. $\overline{AB}$ IS IN P.   (POST. 3)

5. SIMILARLY, BC AND AC ARE IN P, AND THE THEOREM IS PROVED. ∎

THE SYMBOL ∎ IS USED TO INDICATE THAT THE PROOF IS COMPLETE.

Implications

THEOREM 2-1, LIKE ALL THEOREMS, BASICALLY GOES LIKE THIS:

SOMETHING SOMETHING SOMETHING

SOMETHING ELSE.

THE "SOMETHINGS" BETWEEN "IF" AND "THEN" ARE CALLED THE THEOREM'S **HYPOTHESIS** (E.G., "THREE POINTS ARE NONCOLLINEAR").

THE "SOMETHING" AFTER "THEN" IS THE **CONCLUSION** (E.G., "THE LINES DETERMINED BY THREE POINTS LIE IN A PLANE").

IN AN IF-THEN STATEMENT, WE SAY THAT THE HYPOTHESIS **IMPLIES** THE CONCLUSION, AND WE USE THE SYMBOL $\Rightarrow$ FOR "IMPLIES."

HYPOTHESIS $\Rightarrow$ CONCLUSION

IF-THEN STATEMENTS ARE ALSO CALLED **IMPLICATIONS.**

LET ME KNOW IF YOU'RE NOT GETTING THIS, OKAY?

I RESENT THE IMPLICATION...

IT CAN HELP TO THINK OF IMPLICATIONS IN TERMS OF **TERRITORY.** FOR INSTANCE, THE BOROUGH OF BROOKLYN IS A **PART** OF NEW YORK CITY (NYC): IT'S **INSIDE** NYC.

THEN THE STATEMENT **"IF A DOG LIVES IN BROOKLYN, THEN IT LIVES IN NEW YORK CITY"** IS TRUE BECAUSE BROOKLYN IS INSIDE NYC.

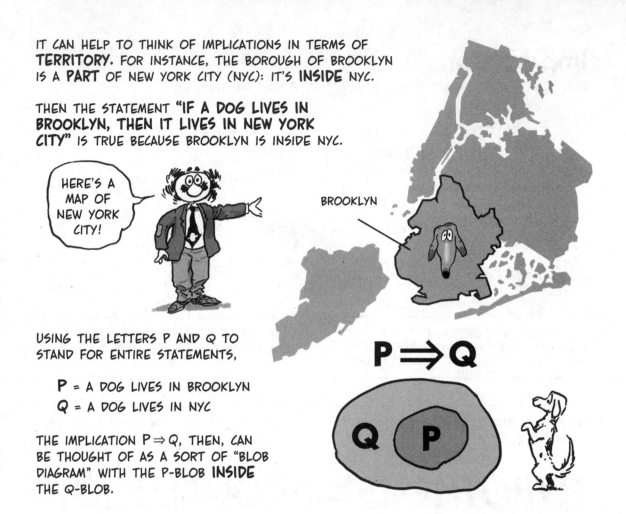

HERE'S A MAP OF NEW YORK CITY!

BROOKLYN

USING THE LETTERS P AND Q TO STAND FOR ENTIRE STATEMENTS,

P = A DOG LIVES IN BROOKLYN

Q = A DOG LIVES IN NYC

THE IMPLICATION P ⇒ Q, THEN, CAN BE THOUGHT OF AS A SORT OF "BLOB DIAGRAM" WITH THE P-BLOB **INSIDE** THE Q-BLOB.

$$P \Rightarrow Q$$

NOW WE MIGHT WONDER ABOUT THE "FLIPPED IMPLICATION," OR **CONVERSE** STATEMENT, Q ⇒ P: "IF A DOG LIVES IN NYC, THEN IT LIVES IN BROOKLYN." AS A GENERAL RULE, THE CONVERSE IS **NOT TRUE.** THE Q-BLOB ISN'T INSIDE THE P-BLOB; THOUSANDS, MAYBE MILLIONS, OF DOGS LIVE IN NYC BUT NOT IN BROOKLYN.

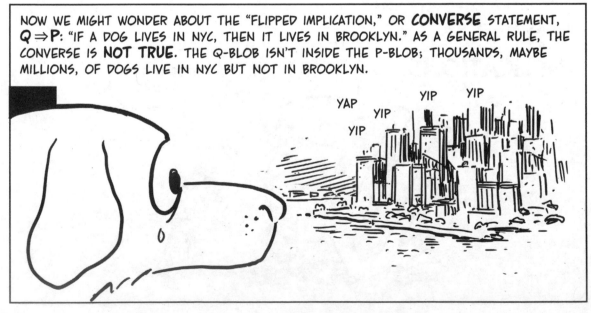

AND YET, SOME STATEMENTS **DO** HAVE A TRUE CONVERSE. FOR EXAMPLE, ON PAGES 66 AND 67 WE'LL PROVE:

IF A TRIANGLE HAS TWO EQUAL SIDES, THEN IT HAS TWO EQUAL ANGLES.

AND CONVERSELY...

IF A TRIANGLE HAS TWO EQUAL ANGLES, THEN IT HAS TWO EQUAL SIDES.

THIS IS A FAIRLY UNUSUAL SITUATION, IN WHICH THE P-BLOB AND THE Q-BLOB **COINCIDE.** EACH ONE IS INSIDE THE OTHER, SO THEY'RE REALLY THE SAME BLOB. TRIANGLES WITH TWO EQUAL SIDES AND TRIANGLES WITH TWO EQUAL ANGLES ARE THE **SAME TRIANGLES.**

P, Q

NOT A POINT'S WORTH OF DIF-FERENCE!

WHEN $P \Rightarrow Q$ AND $Q \Rightarrow P$, WE SAY THAT P AND Q ARE **EQUIVALENT.** "A TRIANGLE HAS TWO EQUAL SIDES" AND "A TRIANGLE HAS TWO EQUAL ANGLES" ARE **TWO WAYS OF SAYING THE SAME THING.** IN THIS CASE, WE SAY "P IF AND ONLY IF Q," OR P IFF Q FOR SHORT, AND EMPLOY THE SYMBOL $\Leftrightarrow$.

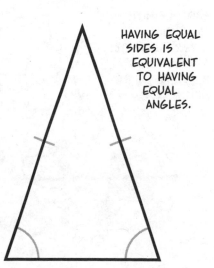

HAVING EQUAL SIDES IS EQUIVALENT TO HAVING EQUAL ANGLES.

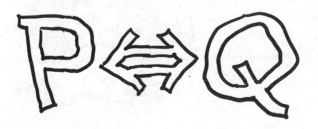

P⟺Q

Definitions

EVERY DEFINITION IS AN IFF STATEMENT. WE DEFINE A TERM BY AN EQUIVALENT DESCRIPTION. FOR INSTANCE, IF WE DEFINE A TRIANGLE AS THREE NONCOLLINEAR POINTS AND THE LINE SEGMENTS BETWEEN THEM, WE ARE REALLY SAYING:

IF A FIGURE CONSISTS OF THREE NONCOLLINEAR POINTS AND THE LINE SEGMENTS BETWEEN THEM, **THEN** IT'S A TRIANGLE,

IF A FIGURE IS A TRIANGLE, **THEN** IT CONSISTS OF THREE NONCOLLINEAR POINTS AND THE LINE SEGMENTS BETWEEN THEM.

SO... NO FOUR-SIDED TRIANGLES, THEN, I GUESS?

VERY SCARCE...

A DEFINITION MUST COMPLETELY CHARACTERIZE THE THING DEFINED. A DEFINITION DESCRIBES ITS SUBJECT AND **NOTHING ELSE.**

BAD DEFINITION: "A HUMAN IS A TWO-LEGGED ANIMAL." BAD BECAUSE WE'RE NOT THE ONLY ONES!

FIGHTING WORDS!

BETTER: "AN ANIMAL IS A HUMAN IFF IT BELONGS TO A SPECIES THAT MANUFACTURES CELL PHONES."

MM-HMM

Indirect Proofs

THE IMPLICATION P⇒Q HAS A DIAGRAM WITH THE P-BLOB INSIDE THE Q-BLOB.

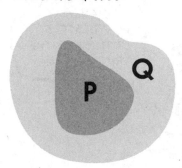

THE REGION **OUTSIDE** A P-BLOB IS CALLED ITS **COMPLEMENT.** IT'S THE BLOB OF THE STATEMENT

NOT-P,

THE **NEGATION** OF P, WRITTEN ∼P: "A DOG DOES **NOT** LIVE IN BROOKLYN."

WHEN Q-BLOB CONTAINS P-BLOB, THE **OUTSIDE** OF P CONTAINS THE **OUTSIDE** OF Q. THAT IS, P⇒Q IS EQUIVALENT TO ∼Q ⇒ ∼P, **NOT-Q IMPLIES NOT-P.**

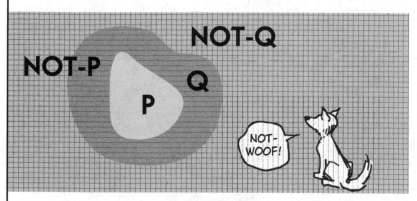

IF A DOG LIVES IN BROOKLYN, THEN IT LIVES IN NEW YORK CITY.

IF A DOG **DOESN'T** LIVE IN NEW YORK CITY, THEN IT **DOESN'T** LIVE IN BROOKLYN.

THE STATEMENT ∼Q ⇒ ∼P IS CALLED THE **CONTRAPOSITIVE** OF P ⇒ Q. YOU'LL FIND SOME GOOD EXAMPLES OF CONTRAPOSITIVES ON PAGE 102 IN THE DISCUSSION OF PARALLEL LINES.

THE CONTRAPOSITIVE LETS US DO **"INDIRECT" PROOFS.** WHEN TRYING TO PROVE P⇒Q, WE SOMETIMES START BY ASSUMING THE **NEGATION OF THE CONCLUSION,** ∼Q, AND SHOW THAT IT IMPLIES ∼P (OR SOME OTHER CONTRADICTION).

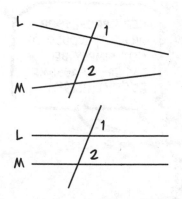

IF THE ANGLES 1 AND 2 ARE UNEQUAL, THEN THE LINES L AND M INTERSECT EACH OTHER.

IF THE LINES L AND M NEVER INTERSECT EACH OTHER, THEN THE ANGLES 1 AND 2 ARE EQUAL.

EXAMPLES TO COME!

HOW MANY POSTULATES ARE NEEDED TO SUPPORT OUR SPECTACULAR STRUCTURE? EUCLID THOUGHT **FIVE** WOULD DO IT, BUT SOME OF HIS EARLIEST READERS QUICKLY FOUND GAPS IN HIS ARGUMENTS. EUCLID HAD MADE UNSTATED ASSUMPTIONS.

BAD GEOMETER!

SIGH... POSTULATE 6: CRITICISM IS THE PRICE OF FAME...

FOR EXAMPLE, GIVEN TWO POINTS ON A LINE, HOW DO WE KNOW THERE ARE ANY POINTS **BETWEEN** THEM? EUCLID'S POSTULATES SAY NOTHING ABOUT IT.

OOPS! I MUSTA BEEN BETWEEN IDEAS...

TO PLUG THESE HOLES, LATER MATHEMATICIANS PILED ON MORE POSTULATES, TWENTY-THREE OF THEM AT ONE POINT, A CLEAR CASE OF CRUELTY TO STUDENTS.

C, B is also between C and A, and there exists a line containing the distinct points A, B, C.

If A and C are two points, then there exists at least one point B on the line AC such that C lies between A and B.

Of any three points situated on a line, there is no more than one which lies between the other two.

Let A, B, C be three points not lying in the same line and let *a* be a line lying in the plane ABC and not passing through any of the points A, B, C.

HERR PROFESSOR HILBERT, WOULD IT NOT SIMPLY BE EASIER TO ASSUME **EVERYTHING?**

26

FINALLY, AROUND 1930(!), GEORGE BIRKHOFF OF HARVARD ACCUSED THE POSTULATE-PUSHERS OF WORKING TOO MUCH IN THE SPIRIT OF EUCLID, WITH THEIR REFUSAL TO USE NUMBERS IN GEOMETRY. **BRING BACK NUMBERS,** SAID BIRKHOFF.

GRAND IDEA, G.B.! NO MATTER WHAT PYTHAGORAS AND EUCLID MAY HAVE THOUGHT ABOUT THEM, NUMBERS ARE OUR FRIENDS. IT PAYS OFF HUGELY TO PUT MEASUREMENT BACK INTO EARTH-MEASUREMENT (EVEN WITHOUT THE EARTH), AND THAT IS EXACTLY WHAT WE WILL DO IN THE NEXT CHAPTER...

Exercises

SOME (BUT NOT ALL!) OF THESE QUESTIONS DON'T EXACTLY HAVE A RIGHT ANSWER, BUT THEY MAY STIMULATE YOUR THINKING.

1a. WE CAN IMAGINE CURVED ONE-DIMENSIONAL THINGS, LIKE CIRCLES OR DOODLES. WE CAN IMAGINE CURVED TWO-DIMENSIONAL THINGS, LIKE SADDLES. CAN WE IMAGINE SOLIDS WITH "INTERNAL" CURVATURE?

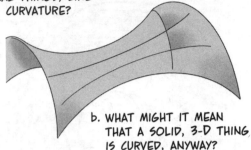

b. WHAT MIGHT IT MEAN THAT A SOLID, 3-D THING IS CURVED, ANYWAY?

2. WHICH POSTULATE IMPLIES THAT EVERY POINT IS ON A LINE? IN SPACE AS YOU IMAGINE IT, IS EVERY POINT ON A LINE? IF NOT, WHAT ABOUT THE LINE BETWEEN THE POINT AND YOUR EYE?

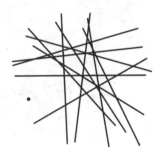

3. SUPPOSE A, B, AND C ARE ON THE SAME LINE. IS THIS THE LINE DETERMINED BY THE TWO POINTS A AND C?

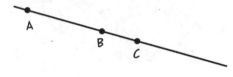

4. SUPPOSE A, B, AND C ARE ON THE SAME LINE L, AND D IS A POINT OFF THE LINE. LET B AND D DETERMINE THE LINE M. HOW MANY TIMES CAN L **INTERSECT** M? THAT IS, HOW MANY POINTS DO L AND M HAVE IN COMMON? WHY?

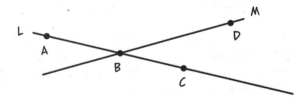

5. IN THE FOLLOWING STATEMENTS, IDENTIFY THE HYPOTHESIS AND THE CONCLUSION.

a. IF A DOG IS WELL TRAINED, THEN IT WILL OBEY ITS OWNER.

b. IF THE SKY IS NOT BLUE, THEN I HAVE LOST MY MIND.

c. IF KLEPTO-MART CHARGES TWO DOLLARS FOR A BAG OF CHIPS, THEN YOU CAN GET A BAG OF CHIPS FOR LESS THAN TWO DOLLARS SOMEWHERE ELSE.

6. WRITE THE CONVERSE AND THE CONTRAPOSITIVE OF EACH STATEMENT IN PROBLEM 5.

7. SUPPOSE P = "A DOG LIVES IN BROOKLYN," AND Q = "A DOG HAS SPOTS." THEIR STATEMENT-BLOBS **OVERLAP**, BECAUSE SOME SPOTTED DOGS LIVE IN BROOKLYN. THIS SHARED REGION CORRESPONDS TO THE STATEMENT **P AND Q**:

"A DOG LIVES IN BROOKLYN **AND** A DOG HAS SPOTS."

WHAT STATEMENT DO YOU THINK GOES WITH THE **COMBINED** BLOB—THE WHOLE SHADED AREA?

Chapter 3
NUMBERS AND LINES

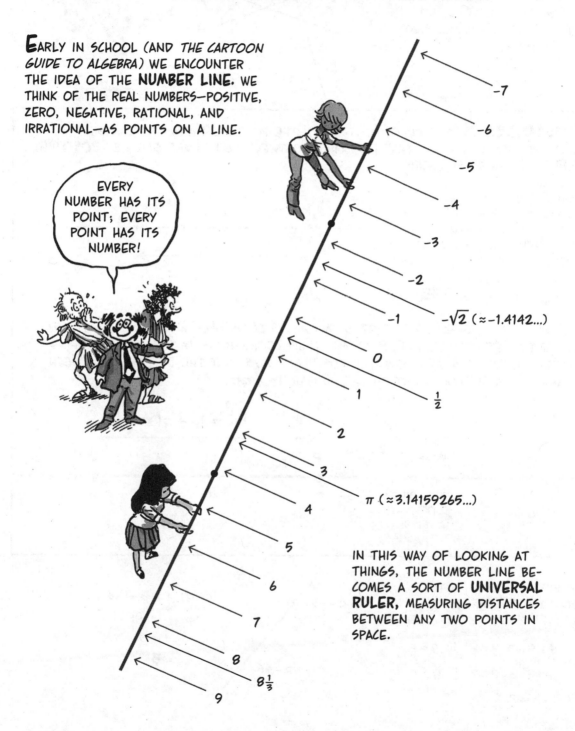

EARLY IN SCHOOL (AND *THE CARTOON GUIDE TO ALGEBRA*) WE ENCOUNTER THE IDEA OF THE **NUMBER LINE.** WE THINK OF THE REAL NUMBERS—POSITIVE, ZERO, NEGATIVE, RATIONAL, AND IRRATIONAL—AS POINTS ON A LINE.

EVERY NUMBER HAS ITS POINT; EVERY POINT HAS ITS NUMBER!

-7
-6
-5
-4
-3
-2
-1
$-\sqrt{2}$ ($\approx$ -1.4142...)
0
1
$\frac{1}{2}$
2
3
π ($\approx$ 3.14159265...)
4
5
6
7
8
$8\frac{1}{3}$
9

IN THIS WAY OF LOOKING AT THINGS, THE NUMBER LINE BE-COMES A SORT OF **UNIVERSAL RULER,** MEASURING DISTANCES BETWEEN ANY TWO POINTS IN SPACE.

IF NUMBERS CORRESPOND TO THE POINTS OF A LINE, THEN **EVERY LINE** SHOULD BE A **NUMBER LINE.** (WE WANT ALL LINES TO BE ALIKE, DON'T WE?) SO LET'S ASSUME THAT TO BE TRUE.

YOU—OUT!

Postulate 5 (THE RULER POSTULATE). THE POINTS ON ANY LINE CAN BE NUMBERED IN SUCH A WAY THAT **DISTANCES** BETWEEN POINTS ARE GIVEN BY **POSITIVE DIFFERENCES** OF NUMBERS.

FOR EXAMPLE, IF THE NUMBERS AT POINTS P AND Q ARE 7.43 AND 2.1, THEN THE DISTANCE **d** FROM P TO Q IS

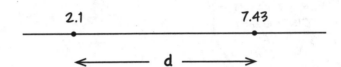

$$d = 7.43 - 2.1 = \mathbf{5.33}$$

THE NUMBER ASSIGNED TO A POINT IS CALLED ITS **COORDINATE.** WE INDICATE POINTS BY CAPITAL LETTERS LIKE A, B, P, Q, AND THEIR COORDINATES BY THE CORRESPONDING LOWERCASE LETTERS a, b, p, q. THE POSTULATE SAYS THAT THE **DISTANCE** BETWEEN P AND Q, OR **PQ** FOR SHORT, IS THE NUMBER $|p - q|$.

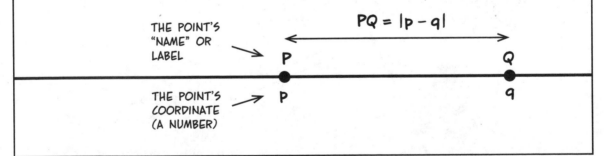

THE POINT'S "NAME" OR LABEL

$$PQ = |p - q|$$

P   Q

THE POINT'S COORDINATE (A NUMBER)

P   q

REMINDER: $|p-q|$ IS THE POSITIVE DIFFERENCE, THE LARGER MINUS THE SMALLER:

$$|p-q| = q-p \quad \text{IF } q \geq p$$

$$|p-q| = p-q \quad \text{IF } q \leq p$$

FOR EXAMPLE,

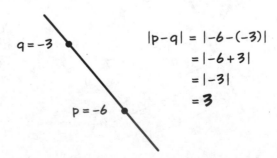

$q = -3$

$p = -6$

$$|p-q| = |-6-(-3)|$$
$$= |-6+3|$$
$$= |-3|$$
$$= \mathbf{3}$$

NOW NUMBERS CONVENIENTLY "LINE UP" IN A DEFINITE **ORDER.** GIVEN TWO NUMBERS, ONE OF THEM IS EITHER LESS THAN, EQUAL TO, OR GREATER THAN THE OTHER.

SO POINTS ON A LINE GO IN ORDER, TOO...

Definition. IF A, B, AND P ARE COLLINEAR POINTS, THEN A AND B ARE **ON THE SAME SIDE** OF P IFF THEIR COORDINATES a AND b ARE (NUMERICALLY) ON THE SAME SIDE OF p, THAT IS, IFF EITHER $p<a$ AND $p<b$ OR $p>a$ AND $p>b$.

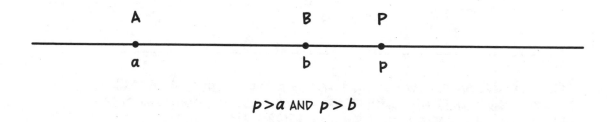

$p>a$ AND $p>b$

Definition. GIVEN A POINT P ON A LINE L, A **RAY ORIGINATING AT P** CONSISTS OF ALL POINTS OF L ON ONE SIDE OF P (INCLUDING P ITSELF).

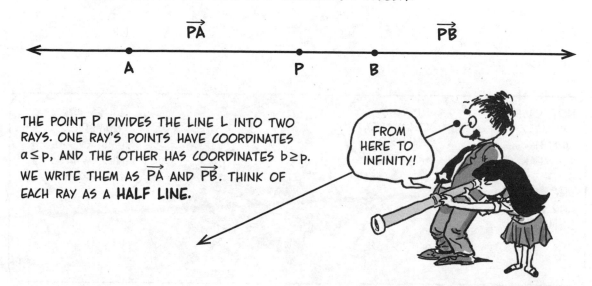

THE POINT P DIVIDES THE LINE L INTO TWO RAYS. ONE RAY'S POINTS HAVE COORDINATES $a \leq p$, AND THE OTHER HAS COORDINATES $b \geq p$. WE WRITE THEM AS $\overrightarrow{PA}$ AND $\overrightarrow{PB}$. THINK OF EACH RAY AS A **HALF LINE.**

FROM HERE TO INFINITY!

31

Definition. THE POINT B IS **BETWEEN** A AND C IFF A, B, AND C ARE **COLLINEAR**, AND A AND C ARE IN **DIFFERENT RAYS** FROM B. TO PUT IT ANOTHER WAY, A AND C ARE **NOT ON THE SAME SIDE** OF B, OR A AND B ARE ON **OPPOSITE SIDES** OF B.

IN TERMS OF COORDINATES, THIS TRANSLATES TO

$$a < b < c$$

OR

$$c < b < a$$

WE SOMETIMES WRITE

A-B-C

TO MEAN THAT B IS BETWEEN A AND C.

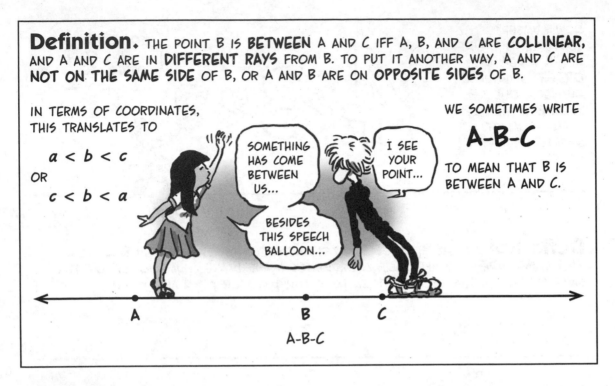

A-B-C

Definition. GIVEN TWO DISTINCT POINTS A AND B, THE **LINE SEGMENT** (OR SIMPLY **SEGMENT**) **AB** CONSISTS OF A, B, AND ALL THE POINTS BETWEEN THEM. A AND B ARE CALLED THE SEGMENT'S **ENDPOINTS**.

A SEGMENT CORRESPONDS TO AN INTERVAL OF NUMBERS. THE POINT **X** IS IN **AB** IFF

$$a \leq x \leq b$$

OR

$$b \leq x \leq a$$

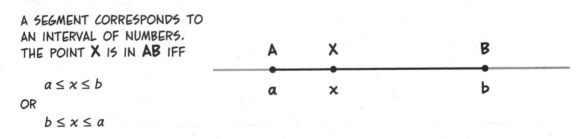

NOTE: WE'RE SNEAKILY WRITING AB TO MEAN BOTH THE **SEGMENT** AND ITS **LENGTH**.

WHEN THEY LIE ALONG THE SAME
LINE AND SHARE AN ENDPOINT,
SEGMENTS CAN BE **ADDED** AND
SUBTRACTED.

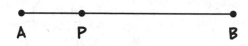

Theorem 3-1. IF A-P-B (P IS BETWEEN A AND B), THEN **AB = AP + PB**.
Proof.

1. A-P-B   (ASSUMED)

2. EITHER $a < p < b$ OR $b < p < a$ (DEF. OF BETWEEN)

3. ASSUME $a < p < b$ (FIRST POSSIBILITY)

4. AP = $p - a$, PB = $b - p$,   (POST. 5)
   AB = $b - a$

5. AP + PB = $(p - a) + (b - p)$ (ADDITION)

6. AP + PB = $(b - a) + (p - p)$
           = $b - a$ (ALGEBRA)

7. AP + PB = AB   (SUBSTITUTING AB FOR $b - a$)

8. IF $b < p < a$, AN IDENTICAL
   ARGUMENT SHOWS THE SAME
   CONCLUSION. ▮

THIS MAY ALL SOUND LIKE MUCH ADO ABOUT
NOT MUCH. WE'VE TAKEN FIVE PAGES TO SHOW
THAT IF A TURTLE WALKS 3 METERS, AND THEN
ANOTHER 4 METERS IN THE SAME DIRECTION, IT
TRAVELS 7 METERS IN ALL. BIG WHOOP!

IT'S A WHOOP
FOR ME!

3   4

AND BIG WHOOP #2: IF THE TURTLE
REVERSES COURSE, WE CAN SUBTRACT, TOO.

Corollary 3-1.1. IF A-P-B, THEN AP = AB − PB

Proof. BY THEOREM 2, IF A-P-B, THEN AP + PB = AB.
SUBTRACTING PB FROM BOTH SIDES GIVES AP = AB − PB. ▮

A   3   P   4   B

THE RULER POS-
TULATE IMPLIES
THAT EVERY LINE
HAS INFINITELY
MANY POINTS,
BECAUSE THERE ARE
INFINITELY MANY
REAL NUMBERS.

IN FACT, THERE ARE INFINITELY MANY POINTS IN ANY **SEGMENT.**

1, 2, 3... OH, YEAH...

AND THIS IN TURN IMPLIES:

Theorem 3-2. THERE ARE INFINITELY MANY LINES THROUGH ANY POINT.

Proof. THE IDEA IS THAT WE CAN CONNECT P TO EVERY POINT ON A LINE.

1. LET P BE A POINT AND L BE ANY LINE NOT CONTAINING P.   (POST. 1)

2. LET A AND B BE TWO DISTINCT POINTS ON L.   (RULER POST.)

3. PA AND PB ARE NOT THE SAME LINE.   (OTHERWISE, P WOULD BE ON L.)

4. IN OTHER WORDS, THERE IS A SEPARATE LINE PX FOR EVERY POINT X IN L.

5. THERE ARE INFINITELY MANY POINTS X IN L, SO THERE ARE INFINITELY MANY LINES PX THROUGH P. ▌

IT'S MUCH EASIER TO DO GEOMETRY WITH NUMBERS, BECAUSE WE'RE ALREADY REASON- ABLY COMFORTABLE WITH NUMBERS.

"WE'RE... **COMFORTABLE"** WITH THEM? IS THAT EVEN **MATH?** I MEAN... **WHAT?**

OKAY! MEASUREMENT IS **BACK**! NOW WE CAN PUT A NUMBER ON THE DISTANCE BETWEEN TWO POINTS!

BUT—

HERE I AM AT POINT A... WAY OVER THERE ARE POINTS B AND C... AND MY UNIVERSAL RULER LETS ME MEASURE AB AND AC.

BUT—

B

C

A

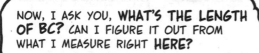

NOW, I ASK YOU, **WHAT'S THE LENGTH OF BC?** CAN I FIGURE IT OUT FROM WHAT I MEASURE RIGHT **HERE?**

B

C

A

BUT—

THAT, BELIEVE IT OR NOT, IS **THE** QUESTION OF BEGINNING GEOMETRY, AND IT'S GOING TO TAKE US **SEVENTEEN MORE CHAPTERS** TO FIND THE ANSWER!

BUT—

BUT **WHAT?**

BUT WHAT ARE THE POSTULATES OF THE **REAL NUMBERS?**

OH! THEM!

AHEM! AS I WAS ABOUT TO SAY... IT'S TIME TO MOVE ON TO MORE **EXCITING GEOMETRY!!**

CAN BEING KICKED DOWN THE ROAD

Exercises

1. HERE IS A LINE WITH COORDINATES ASSIGNED TO ITS POINTS. INDICATE (APPROXIMATELY) ON THE LINE WHERE YOU WOULD FIND POINTS WITH THESE COORDINATES:

a. 3   b. –3   c. –4   d. $3\frac{1}{2}$ e. $-\frac{1}{2}$ f. 21/5   g. 2.25   h. –15/4   i. 0.375   j. 5.01

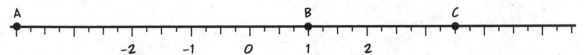

2. ON THE SAME LINE, WHAT IS THE LENGTH OF AB? WHAT IS THE LENGTH OF BC? OF AC?

3. HERE WE PUT DOWN COORDINATES ON A LINE IN TWO DIFFERENT WAYS. CALCULATE THE LENGTH OF AB USING EACH SET OF COORDINATES. ARE THE ANSWERS THE SAME OR DIFFERENT? WHY?

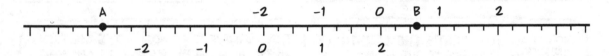

4a. IS –5<–3 OR –5>–3?   b. WHAT IS –3–(–5)?   c. WHAT IS –5–(–3)?   d. WHAT IS |–5–(–3)|?

5. IN EACH DIAGRAM, IS A-B-C?

a.
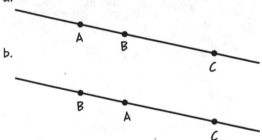
b.

c.
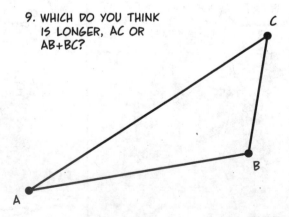

d.

6. IF A-B-C AND B-C-D, IS A-B-D?

7. IF A-B-D AND A-C-D, SHOW THAT BC<AD.

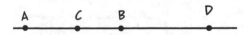

8. A RULE OF INEQUALITIES IS THAT IF a<b AND c IS ANY NUMBER, THEN a+c<b+c AND a–c<b–c.

USE THIS TO SHOW THAT b>a ⇔ b–a>0.

9. WHICH DO YOU THINK IS LONGER, AC OR AB+BC?

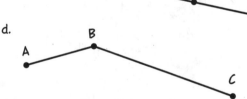

Chapter 4
COMPASS AND CIRCLE

KEVIN NEEDS TO DO SOME PRACTICAL GEOMETRY. HE WANTS TO REPLACE A MISSING BOARD, SO HE NEEDS TO CUT A NEW ONE TO FIT.

ANY NORMAL CARPENTER WOULD FIND THE LENGTH WITH A TAPE MEASURE, BUT KEVIN IS FEELING EUCLIDEAN. ALL HE HAS IS A PIN, SOME STRING, AND A PENCIL.

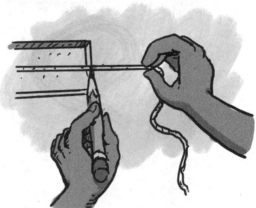

AND A SAW!

HE PINS THE STRING TO ONE END OF THE SPACE WHERE THE BOARD WILL GO...

PULLS THE STRING TAUT AND MARKS IT AT THE OTHER END OF THE GAP...

UNPINS THE STRING, STICKS THE PIN TO THE END OF A LONG BOARD, PULLS THE STRING TAUT, AND MARKS THE BOARD WHERE THE STRING'S MARK TOUCHES IT.

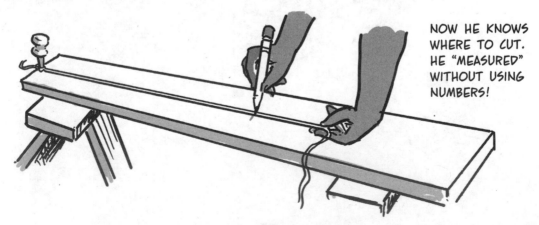

NOW HE KNOWS WHERE TO CUT. HE "MEASURED" WITHOUT USING NUMBERS!

IN FACT, KEVIN HAS
MADE A SIMPLE
COMPASS. IF HE PINS
A POINT OF THAT
STRING TO THE FLOOR,
HE CAN TIE A PENCIL
SOMEWHERE ELSE ON
THE STRING AND DRAW
A CIRCLE. SO LET'S
TALK ABOUT CIRCLES.

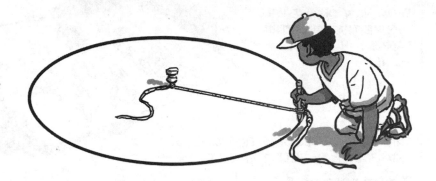

Definition. A CIRCLE CONSISTS OF ALL POINTS IN A PLANE AT THE SAME DISTANCE FROM A POINT CALLED ITS **CENTER.** GIVEN A POINT C IN A PLANE P, AND A POSITIVE NUMBER r, THEN A POINT B IN THE PLANE P IS ON **THE CIRCLE WITH CENTER C AND RADIUS r** IFF CB=r.

A LINE SEGMENT CB
WITH ONE ENDPOINT AT
THE CENTER AND THE
OTHER ON THE CIRCLE
IS CALLED A **RADIUS.**

THE WORD "RADIUS"
CAN ALSO MEAN r,
THE LENGTH OF CB.
ALL RADII (THE PLURAL
OF "RADIUS") HAVE
THE SAME LENGTH,
BY DEFINITION.

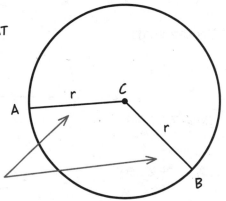

RADII

WHEN A SEGMENT AB BETWEEN TWO POINTS ON THE CIRCLE ALSO CONTAINS THE CENTER, THE SEGMENT AND ITS LENGTH ARE CALLED A **DIAMETER.** IN A DIAMETER, A-C-B. IF THE RADIUS IS r, THEN

$$AB = AC + CB \quad \text{(THM. 3-1)}$$
$$= r + r$$
$$= 2r$$

THE DIAMETER IS TWICE
THE RADIUS.

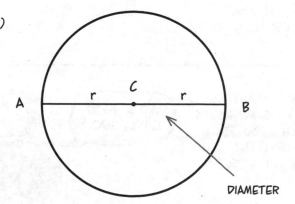

DIAMETER

FROM NOW ON, WE WILL ASSUME THAT EVERYTHING HAPPENS IN A SINGLE PLANE.

WELL, THAT'S **OUR** REALITY, ISN'T IT?

THIS NEXT RESULT SAYS THAT CIRCLES AND LINES PLAY WELL TOGETHER.

Theorem 4-1. GIVEN A CIRCLE WITH CENTER C AND RADIUS r, IF ANY LINE (IN THE SAME PLANE!) CONTAINS C, THEN THE LINE INTERSECTS THE CIRCLE TWICE.

Proof.

1. ASSIGN COORDINATES TO THE LINE L, SO THAT C HAS COORDINATE c.   (RULER POST.)

2. LET P BE THE POINT ON L WITH COORDINATE $c-r$, AND Q BE THE POINT WITH COORDINATE $c+r$.   (RULER POST.)

3. $PC = c - (c-r) = r$
   $CQ = (c+r) - c = r$ (RULER POST.)

4. SO BOTH P AND Q LIE ON THE CIRCLE.   (DEF. OF CIRCLE)

5. THE LINE CAN'T INTERSECT MORE THAN TWICE. WHY? ANY POINT OF INTERSEC-TION WOULD NECESSARILY HAVE COORDINATE EITHER $c+r$ OR $c-r$, AND SO WOULD BE IDENTICAL TO P OR Q. ∎   (RULER POST.)

L — P ($c-r$) — C (c) — Q ($c+r$)

$PC = CQ = r$

THIS THEOREM GUARANTEES THAT CIRCLES HAVE NO HOLES THROUGH WHICH A RAY FROM THE CENTER MIGHT SLIP WITH-OUT CROSSING THE CIRCLE.

AND A GOOD THING, TOO.

THEOREM 4-1 SAYS IT'S OKAY TO USE A COMPASS TO ADD AND SUBTRACT LENGTHS.

WELL, I KNEW **THAT!!**

LET'S ADD TWO SEGMENTS AB AND PQ (**NOT** NECESSARILY ON THE SAME LINE) WITH LENGTHS r AND s RESPECTIVELY.

WE'LL ADD IT HERE...

THE SEGMENT AB IS PART OF AN ENTIRE LINE $\overleftrightarrow{AB}$. NUMBER THE LINE'S POINTS. WE MAY ASSUME $a < b$.

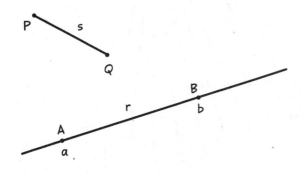

DRAW A CIRCLE OF RADIUS s CENTERED AT B. THE CIRCLE INTERSECTS THE LINE TWICE, AT C AND D, BY THEOREM 4-1.

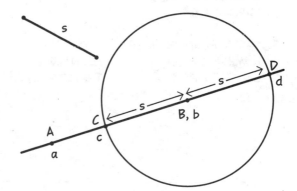

THE COORDINATE d OF D IS b + s, BY THE RULER POSTULATE, SO

$$AD = AB + BD$$
$$= r + s$$
$$= AB + PQ$$

AND WE HAVE SUCCESSFULLY ADDED AB TO PQ.

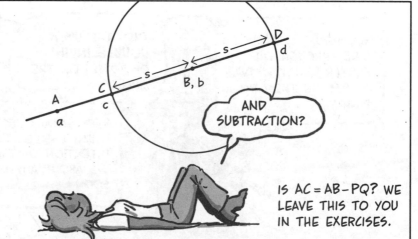

AND SUBTRACTION?

IS AC = AB − PQ? WE LEAVE THIS TO YOU IN THE EXERCISES.

Theorem 4-2. EVERY CIRCLE HAS INFINITELY MANY POINTS.

Proof.

1. LET C BE THE CENTER.   (DEF. OF CIRCLE)

2. EACH LINE THROUGH C INTERSECTS THE CIRCLE TWICE.   (THM. 4-1)

3. DIFFERENT LINES THROUGH C INTERSECT THE CIRCLE IN DIFFERENT PAIRS OF POINTS.   (OTHERWISE THEY WOULD BE THE SAME LINE, BY POST. 2.)

4. THERE IS AN INFINITE NUMBER OF LINES THROUGH C.   (THM. 3-2)

5. THERE IS AN INFINITE NUMBER OF POINTS ON THE CIRCLE. ∎   (THEY ARE COUNTED BY THE LINES.)

WE'LL HAVE MORE TO SAY ABOUT THIS IN THE NEXT CHAPTER.

SO, ACTUALLY, EACH LINE THROUGH THE CENTER "COUNTS" TWO POINTS ON THE CIRCLE, RIGHT?

YES.

THEN THERE'S A **DOUBLE INFINITY** OF POINTS ON THE CIRCLE?

VERY GOOD QUESTION, WHICH I HAVE ZERO INTENTION OF TALKING ABOUT ANYTIME SOON...

BECAUSE CIRCLES CURVE AROUND, A PAIR OF CIRCLES MAY INTERSECT EACH OTHER AT AS MANY AS TWO POINTS, UNLIKE LINES, WHICH CAN INTERSECT ONLY ONCE, BY POSTULATE 2.

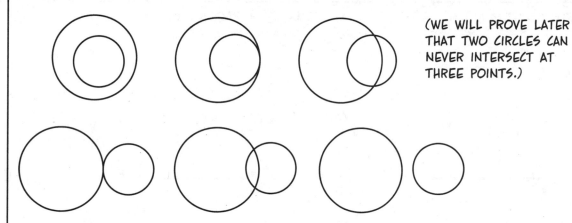

(WE WILL PROVE LATER THAT TWO CIRCLES CAN NEVER INTERSECT AT THREE POINTS.)

TWO CIRCLES INTERSECT TWICE, ROUGHLY SPEAKING, WHEN THEY ARE LARGE ENOUGH AND CLOSE ENOUGH TOGETHER. LET'S MAKE THIS PRECISE.

SUPPOSE CIRCLE C_A HAS CENTER **A** AND RADIUS r_A, AND CIRCLE C_B HAS CENTER **B** AND RADIUS r_B. ASSUME THAT **A** IS ON OR OUTSIDE C_B AND **B** IS ON OR OUTSIDE C_A, THAT IS

$$AB \geq r_A, \ AB \geq r_B$$

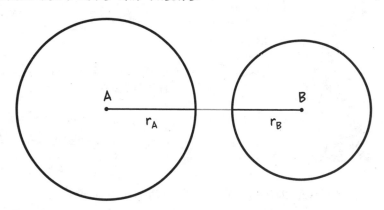

Postulate 6. GIVEN CIRCLES AS JUST DESCRIBED, IF $AB < r_A + r_B$, THEN THE CIRCLES INTERSECT AT TWO POINTS.

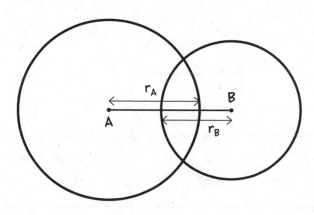

THIS POSTULATE IS ALL-IMPORTANT WHEN MAKING GEOMETRICAL DRAWINGS, OR **CONSTRUCTIONS**. HERE'S THE FIRST OF MANY EXAMPLES TO COME.

START WITH ANY LINE SEGMENT AB OF LENGTH r.

DRAW CIRCLE C_A OF RADIUS r CENTERED AT A AND CIRCLE C_B OF RADIUS r CENTERED AT B.

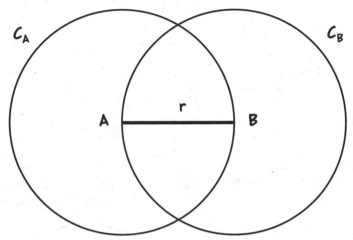

THIS SEGMENT LOOKS GOOD ...

THESE CIRCLES SATISFY THE CONDITIONS OF POSTULATE 6.

$$r_A = r_B = r$$
$$r = AB < AB + AB = 2r$$

SO THE CIRCLES INTERSECT AT TWO POINTS, D AND E.

AD IS A RADIUS OF THE CIRCLE C_A, SO

AD = AB

BD IS A RADIUS OF THE CIRCLE C_B, SO

BD = AB

AND WE HAVE JUST MADE A TRIANGLE (TWO TRIANGLES, ACTUALLY) WITH **ALL SIDES EQUAL TO THE GIVEN SEGMENT.**

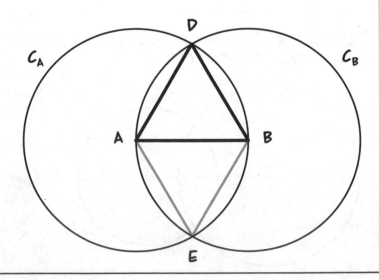

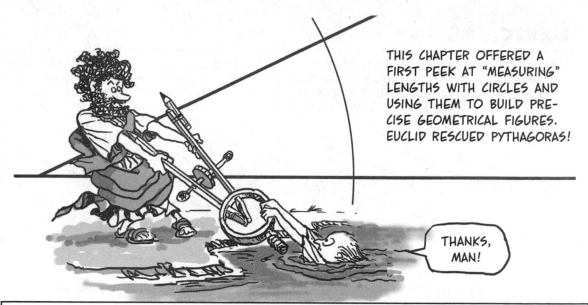

THIS CHAPTER OFFERED A FIRST PEEK AT "MEASURING" LENGTHS WITH CIRCLES AND USING THEM TO BUILD PRECISE GEOMETRICAL FIGURES. EUCLID RESCUED PYTHAGORAS!

THANKS, MAN!

ON THE OTHER HAND, EUCLID WOULD SURELY DISAPPROVE OF OUR NEXT CHAPTER, WHICH EXPLAINS HOW TO USE CIRCLES FOR MEASURING SOMETHING ELSE ENTIRELY...

HM, WONDER WHAT THAT WOULD BE...

HOW SHOULD I KNOW?

Exercises

1. GIVEN SEGMENT AB, HOW CAN YOU EXTEND IT TO A SEGMENT WITH THREE TIMES THE LENGTH? FOUR TIMES? CAN YOU THINK OF A WAY TO MAKE A SEGMENT EIGHT TIMES AS LONG WITH JUST THREE SWEEPS OF THE COMPASS?

2. IN THIS DIAGRAM, IF PQ=55.3 AND PC=88.8, WHAT'S THE RADIUS OF THE CIRCLE? ASSIGN COORDINATES ON THE SEGMENT PC CONSISTENT WITH THOSE LENGTHS.

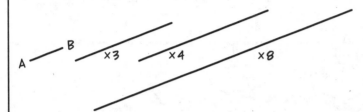

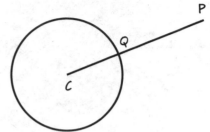

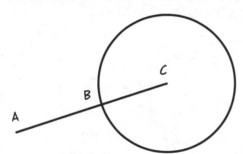

3. GIVEN A CIRCLE CENTERED AT C, ANY POINT A OUTSIDE THE CIRCLE, AND AC INTERSECTS THE CIRCLE AT B, SHOW THAT AB=AC−BC.

4. IS THIS DIAGRAM POSSIBLE AS DRAWN?

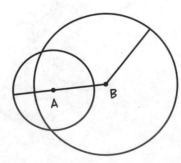

AB=7

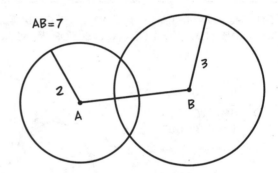

5. WHAT HAPPENS TO POSTULATE 6 WHEN ONE CIRCLE'S CENTER IS INSIDE ANOTHER CIRCLE? LET A AND B BE THE TWO CENTERS, r_A AND r_B THE RADII. ASSUME THAT CENTER A IS INSIDE CIRCLE B AND THAT $r_A < r_B$.

a. WHAT INEQUALITY DESCRIBES THE SITUATION WHEN CIRCLE A IS TOO SMALL TO INTERSECT CIRCLE B?

b. WHAT INEQUALITY DESCRIBES THE SITUATION WHEN CIRCLE A DOES INTERSECT CIRCLE B?

c. HOW ABOUT WHEN THE TWO CIRCLES TOUCH AT A SINGLE POINT?

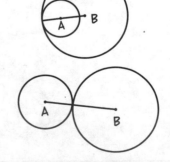

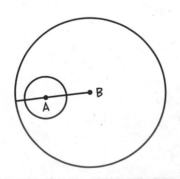

Chapter 5
ANGLES

WHERE DO RAYS GO WHEN THEY GO?

BIANCA IS WATCHING TRAILS OF ANTS AS THEY ENTER AND LEAVE THEIR HOME, A HOLE IN THE GROUND. THESE ARE VERY SPECIAL ANTS THAT MARCH IN PERFECTLY STRAIGHT LINES. ARMY ANTS, MAYBE, WITH TINY DRILL SERGEANTS.

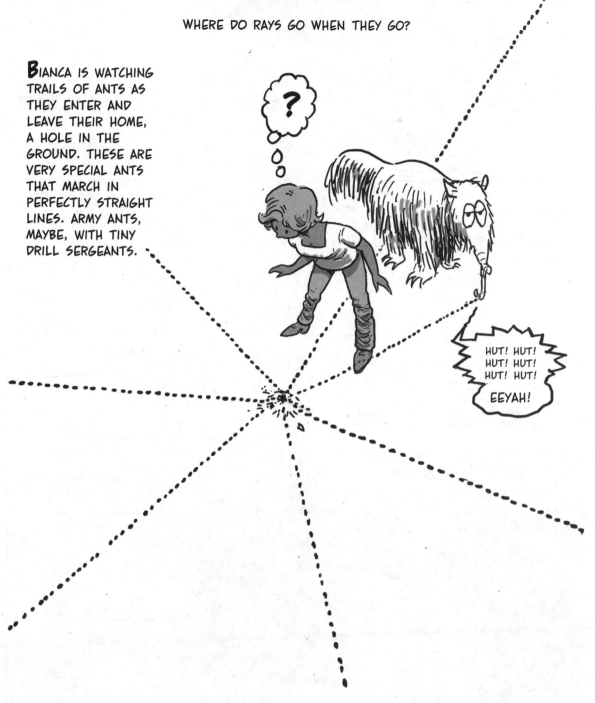

HUT! HUT! HUT! HUT! HUT! HUT!

EEYAH!

BIANCA WANTS A WAY TO DESCRIBE WHERE THESE INSECT PLATOONS ARE GOING.

THIS IS DIFFERENT FROM DISTANCE! TWO ANTS ON SEPARATE LINES CAN BE EQUALLY FAR FROM THE HOLE, AND TWO ANTS ON THE SAME LINE ARE AT DIFFERENT DISTANCES FROM THE HOLE.

MARCH, YOU OVERGROWN LARVAE!

SAME DISTANCE FROM HOME, ON DIFFERENT LINES

ON SAME LINE BUT DIFFERENT DISTANCES FROM HOME

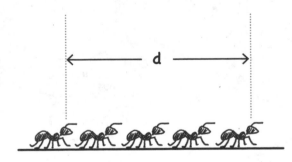

WE MIGHT THINK OF IT THIS WAY. DISTANCE MEASURES THE SEPARATION BETWEEN **TWO POINTS** (OR ANTS) **ON A SINGLE LINE.**

d

WHAT BIANCA WANTS TO MEASURE IS THE SPREAD BETWEEN **TWO LINES THROUGH THE SAME POINT.**

OR TWO RAYS, REALLY...

SO WHAT SHOULD BIANCA DO?

DEFINE HER TERMS, NATURALLY!

OKAY, THIS TIME WE AGREE...

Definition. TWO RAYS FORM AN **ANGLE** IFF THEY ORIGINATE AT THE SAME POINT. THIS POINT (O IN THE DIAGRAM) IS CALLED THE **VERTEX** OF THE ANGLE.

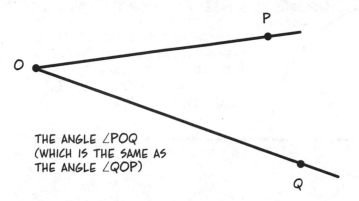

THE ANGLE ∠POQ (WHICH IS THE SAME AS THE ANGLE ∠QOP)

WE SOMETIMES INDI-CATE AN ANGLE SIMPLY AS ∠O. WE MAY ALSO WRITE ∠POQ, IF WE HAVE SOME SPECIAL INTEREST IN POINTS P AND Q ON THE TWO RAYS.

YOU CAN'T MEASURE AN ANGLE WITH A RULER BECAUSE ANGLES SPREAD OUT AS YOU GO FARTHER FROM THE VERTEX. WHERE WOULD YOU MEASURE?

UM...

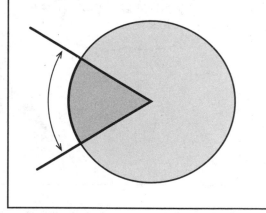

INSTEAD WE MEASURE AN ANGLE WITH A **CIRCLE**. WE ASK: **WHAT PART OF A CIRCLE** DOES THE ANGLE "EAT UP"?

WHICH CIRCLE? **ANY** CIRCLE CENTERED AT THE VERTEX WILL DO; THE ANGLE MARKS OFF THE SAME FRACTION OF A BIG CIRCLE AS A SMALL ONE. AS LONG AS THE CIRCLES ARE CENTERED AT THE VERTEX, IT DOESN'T MATTER WHICH ONE YOU USE.

YES!!

49

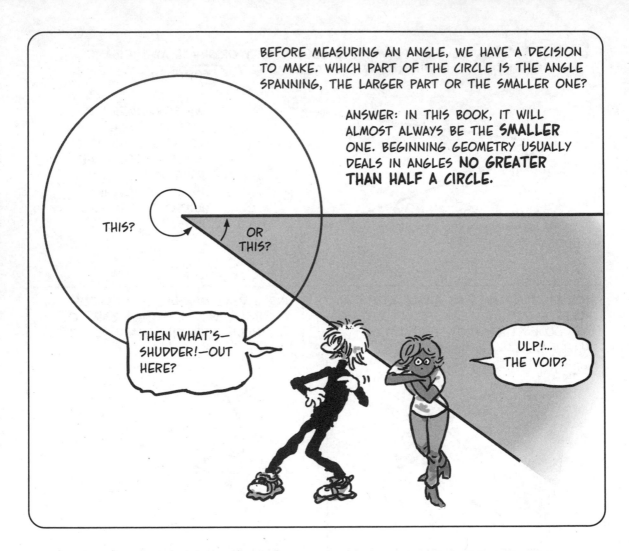

BEFORE MEASURING AN ANGLE, WE HAVE A DECISION TO MAKE. WHICH PART OF THE CIRCLE IS THE ANGLE SPANNING, THE LARGER PART OR THE SMALLER ONE?

ANSWER: IN THIS BOOK, IT WILL ALMOST ALWAYS BE THE **SMALLER** ONE. BEGINNING GEOMETRY USUALLY DEALS IN ANGLES **NO GREATER THAN HALF A CIRCLE.**

THIS?

OR THIS?

THEN WHAT'S— SHUDDER!—OUT HERE?

ULP!... THE VOID?

FROM THE BABYLONIANS* WE BORROW THE IDEA OF DIVIDING THE CIRCLE INTO 360 EQUAL PARTS, OR **DEGREES.** A DEGREE HAS THE SYMBOL °, AS IN **45°**, 45 DEGREES. A HALF-CIRCLE, THEN, HAS 180°.

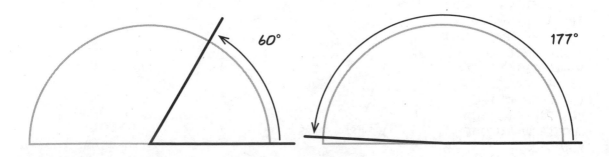

60°

177°

*SEE PAGE 4.

50

SO WE'LL MEASURE ANGLES IN DEGREES, TOO, FROM 0° TO 180°, ACCORDING TO HOW MUCH CIRCLE THEY SWEEP OUT. IN EFFECT, WE'RE WRAPPING A FLEXIBLE RULER, LIKE A TAILOR'S TAPE MEASURE, AROUND A SEMICIRLE.

NOTE: AGAIN WE'RE PARTING WAYS WITH EUCLID, WHO NEVER MEASURED ANYTHING.

THE RESULTING DEVICE, A **PROTRACTOR**, IS A SEMICIRCULAR SCALE USUALLY MADE OF PLASTIC, WOOD, OR METAL, WITH DEGREES GOING IN EQUAL STEPS FROM 0° ON THE RIGHT TO 180° ON THE LEFT. THE BASE, A DIAMETER, HAS ITS CENTER CLEARLY MARKED.

To Measure an Angle

FIRST PLACE THE ANGLE'S VERTEX AT THE CENTER OF THE PROTRACTOR'S BASE SO THAT THE RAYS FALL ANYWHERE WITHIN THE SEMICIRCLE;

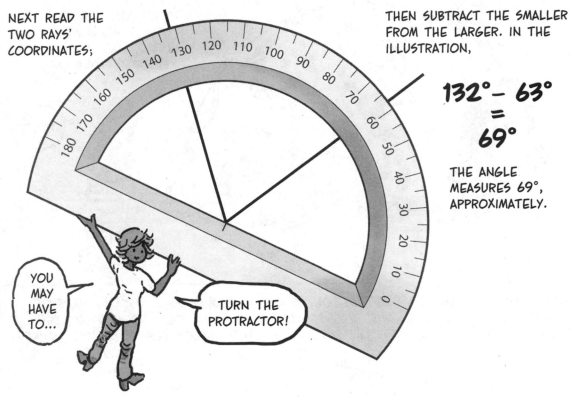

NEXT READ THE TWO RAYS' COORDINATES;

THEN SUBTRACT THE SMALLER FROM THE LARGER. IN THE ILLUSTRATION,

$$132° - 63° = 69°$$

THE ANGLE MEASURES 69°, APPROXIMATELY.

YOU MAY HAVE TO...

TURN THE PROTRACTOR!

ANY ANGLE SMALLER THAN 90° IS CALLED **ACUTE.**

ANYTHING GREATER THAN 90° IS **OBTUSE.**

OBTUSE MEANS THERE'S ROOM TO CUT LOOSE!

ACUTE LIKE THE TOE OF MY BOOT...

WE NOW POSTULATE THE MEASUREMENT OF ANGLES BY PROTRACTORS, MUCH AS WE MEASURE DISTANCE WITH RULERS. HERE, THOUGH, WE NEED TO TAKE EXTRA CARE TO INCLUDE ONLY ANGLES NOT MORE THAN 180°.

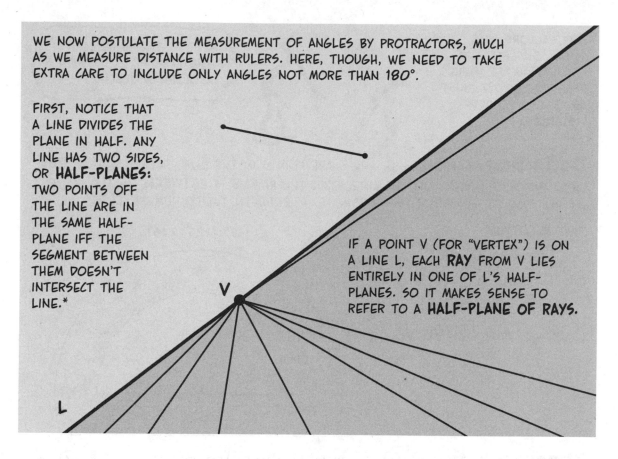

FIRST, NOTICE THAT A LINE DIVIDES THE PLANE IN HALF. ANY LINE HAS TWO SIDES, OR **HALF-PLANES:** TWO POINTS OFF THE LINE ARE IN THE SAME HALF-PLANE IFF THE SEGMENT BETWEEN THEM DOESN'T INTERSECT THE LINE.*

IF A POINT V (FOR "VERTEX") IS ON A LINE L, EACH **RAY** FROM V LIES ENTIRELY IN ONE OF L'S HALF-PLANES. SO IT MAKES SENSE TO REFER TO A **HALF-PLANE OF RAYS.**

AND SO...

Postulate 7 (THE PROTRACTOR POSTULATE). GIVEN A HALF-PLANE OF RAYS FROM A VERTEX V, THE RAYS CAN BE NUMBERED FROM 0 TO 180 SO THAT POSITIVE DIFFERENCES BETWEEN NUMBERS MEASURE ANGLES.

WRITING m(VA) FOR THE PROTRACTOR COORDINATE OF RAY VA, THE POSTULATE SAYS THAT

$$\angle AVB = |m(VA) - m(VB)|$$

∠AVB MEANS BOTH THE ANGLE AND ITS SIZE IN DEGREES. FOR INSTANCE, THE ANGLE IN BOLD MEASURES

$$\angle AVB = 125° - 45° = 80°$$

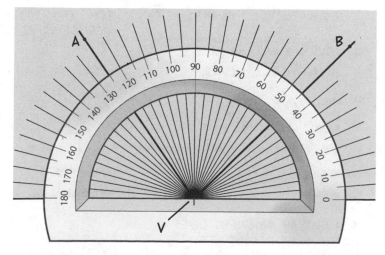

*THERE IS A HIDDEN ASSUMPTION HERE—SORRY!

PROTRACTORS MAKE SOME CONCEPTS AND RESULTS ABOUT ANGLES EASIER TO PROVE, JUST AS RULERS DO WITH DISTANCES.

AND AT THE SAME COST: THE PAGE FILLS UP WITH ALGEBRA SCRIBBLES!

Definition. SUPPOSE A, B, AND C ARE POINTS IN THE SAME HALF-PLANE OF SOME LINE, AND V IS A POINT ON THE LINE. THEN THE RAY $\vec{VB}$ IS **BETWEEN** $\vec{VA}$ AND $\vec{VC}$ IFF ITS PROTRACTOR MEASURE (ITS COORDINATE) IS BETWEEN THEIRS NUMERICALLY.

THAT IS, EITHER

$$m(VA) > m(VB) > m(VC)$$

OR

$$m(VA) < m(VB) < m(VC)$$

IF SO, WE WRITE $\vec{VA}$-$\vec{VB}$-$\vec{VC}$.

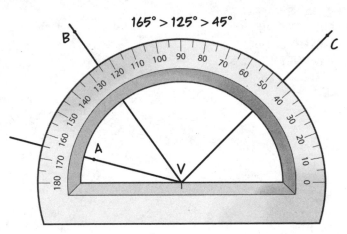

$$165° > 125° > 45°$$

Theorem 5-1. IF $\vec{VA}$-$\vec{VB}$-$\vec{VC}$, THEN THE ANGLES ADD UP: $\angle CVA = \angle CVB + \angle BVA$

Proof. THE PROOF IS EXACTLY LIKE THE PROOF OF THEOREM 3-1 ON PAGE 33: EACH ANGLE IS A DIFFERENCE OF TWO COORDINATES. ADDING THE DIFFERENCES AND CANCELING THE MIDDLE TERM GIVES THE RESULT.

1. ASSUME $m(\vec{VC}) > m(\vec{VB}) > m(\vec{VA})$ (ONE POSSIBILITY, BY DEF. OF VA-VB-VC)

2. $\angle CVB = m(\vec{VC}) - m(\vec{VB})$
   $\angle BVA = m(\vec{VB}) - m(\vec{VA})$ (POST. 7)

3. THEN $\angle CVB + \angle BVA$
   $= (m(\vec{VC}) - m(\vec{VB})) + (m(\vec{VB}) - m(\vec{VA}))$
   $= m(\vec{VC}) - m(\vec{VA}) + (m(\vec{VB}) - m(\vec{VB}))$
   $= m(\vec{VC}) - m(\vec{VA})$ (ALGEBRA)
   $= \angle CVA$ (POST. 7)

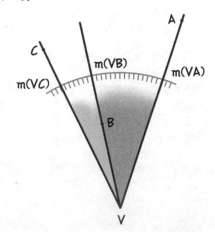

4. A SIMILAR ARGUMENT GIVES THE SAME RESULT WHEN $m(VC) < m(VB) < m(VA)$. ∎

IF TWO OF THE THREE RAYS FROM A VERTEX V ARE OPPOSITE RAYS ON THE SAME LINE, THE TWO ADJACENT ANGLES ARE CALLED A **LINEAR PAIR.** THE ANGLES IN A LINEAR PAIR ADD TO 180°.

$$\angle AVB + \angle BVC = 180°$$

BY THEOREM 5-1.

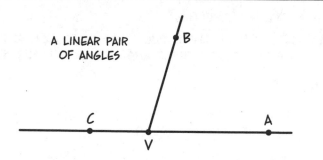

A LINEAR PAIR OF ANGLES

WHEN TWO LINES CROSS, EACH PAIR OF **ADJACENT** ANGLES (LIKE ∠2 AND ∠4) FORMS A LINEAR PAIR. THE ANGLES **OPPOSITE** EACH OTHER (∠1 AND ∠4, ∠2 AND ∠3) ARE KNOWN AS **VERTICAL ANGLES,*** AND THEY HAVE A VERY SPECIAL RELATIONSHIP.

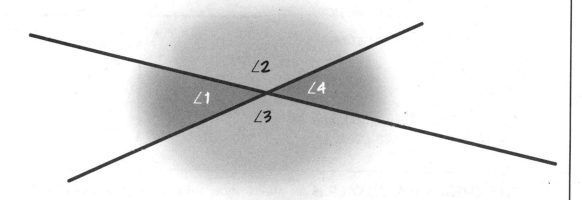

Theorem 5-2. VERTICAL ANGLES ARE EQUAL.
Proof.

1. ∠1 AND ∠2 ARE A LINEAR PAIR.   (DEF.)
2. ∠2 AND ∠4 ARE A LINEAR PAIR.   (DEF.)
3. ∠1 + ∠2 = 180°, ∠2 + ∠4 = 180°   (THM. 6-2)
4. ∠1 + ∠2 = ∠2 + ∠4   (SUBSTITUTION)
5. ∠1 = ∠4   (SUBTRACTION)
6. SIMILARLY, ∠2 = ∠3. ∎

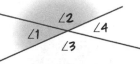

THE VERTICAL ANGLE THEOREM WILL PLAY A ROLE IN COUNTLESS THEOREMS. IT ALWAYS PAYS TO LOOK FOR PAIRS OF VERTICAL ANGLES!

THEY MUST BE SOMEWHERE...

*BECAUSE THEY LIE ACROSS A **VERTEX,** NOT BECAUSE THEY GO UP AND DOWN.

THIS FOLLOWS DIRECTLY:

Theorem 5-3. IF TWO LINES INTERSECT, AND ONE ANGLE MEASURES 90°, THEN EACH OF THE OTHER THREE ANGLES ALSO MEASURES 90°.

Proof.

1. ∠1 AND ∠2 ARE A LINEAR PAIR.   (DEF. OF LINEAR PAIR)

2. ∠1 + ∠2 = 180°   (THM. 5-1)

3. ∠1 = 90°   (ASSUMED)

4. 90° + ∠2 = 180°   (SUBSTITUTION)

5. ∠2 = 90°   (SUBTRACTION)

6. ∠1 AND ∠4 ARE VERTICAL ANGLES. ∠2 AND ∠3 ARE VERTICAL ANGLES.   (DEFINITION)

7. ∠4 = ∠1 = 90° ∠3 = ∠2 = 90° ▌   (VERTICAL ANGLES ARE EQUAL.)

Definitions. A RIGHT ANGLE IS AN ANGLE MEASURING 90°. TWO LINES THAT MEET IN A RIGHT ANGLE ARE CALLED PERPENDICULAR. WE WRITE L⊥M FOR "L IS PERPENDICULAR TO M." WE ALSO INDICATE RIGHT ANGLES WITH A LITTLE SQUARE AT THE VERTEX.

More Terminology and Notation

WHEN TWO ANGLES IN A DIAGRAM ARE EQUAL, WE OFTEN DRAW A LITTLE ARC IN THEM TO INDICATE EQUALITY.

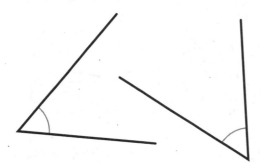

WE MAY USE MULTIPLE ARCS WHEN SEVERAL ANGLES ARE PRESENT.

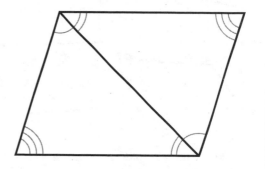

TWO ANGLES THAT ADD TO **90°** ARE CALLED **COMPLEMENTARY**. THEY MAY OR MAY NOT BE ADJACENT.

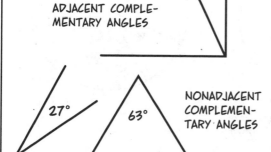

ADJACENT COMPLE-MENTARY ANGLES

27°

63°

NONADJACENT COMPLEMEN-TARY ANGLES

TWO ANGLES THAT ADD TO **180°** ARE CALLED **SUPPLEMENTARY**. EVERY LINEAR PAIR IS SUPPLEMENTARY, BUT SUPPLEMEN-TARY ANGLES NEED NOT BE A LINEAR PAIR.

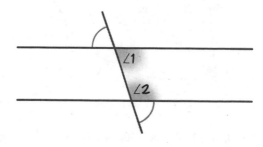

∠1

∠2

IN THE EXERCISES, YOU WILL PROVE THAT ANGLES ∠1 AND ∠2 ARE SUPPLEMENTARY.

THIS BRINGS AN END TO OUR TOUR OF GEOMETRY'S BASIC INGREDIENTS: LINES, RAYS, SEGMENTS, CIRCLES, ARCS, AND ANGLES.

NOW WE'RE READY TO TALK ABOUT THEIR **COMBINATIONS**, STARTING WITH THE VERY SIMPLEST, THE **TRIANGLE**.

OOH! TOUGH!

Exercises

1. IS THE ANGLE ACUTE OR OBTUSE?

a.

b.

c.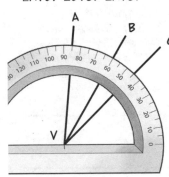

d.

2. WHAT'S THE MEASURE OF ∠AVB?

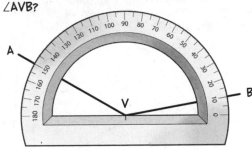

3. WHAT'S THE MEASURE OF ∠AVB? ∠BVC? ∠AVC?

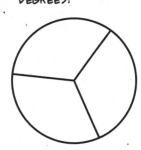

4. IF THESE THREE ANGLES ARE EQUAL, WHAT ARE THEY IN DEGREES?

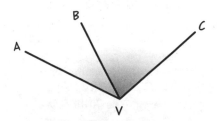

5. WHAT'S THE MEASURE IN DEGREES OF THE SHADED ANGLE?

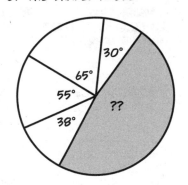

30°
65°
55°
??
38°

6. PROVE THAT ANGLES SUBTRACT AS WELL AS ADD: IF AV-BV-CV, SHOW THAT ∠BVC = ∠AVC − ∠AVB.

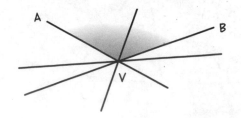

7. IF ∠AVB=27°, WHAT'S ∠CVE?

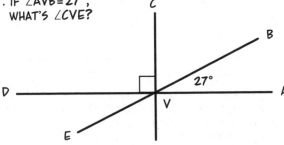

27°

8. INDICATE TWO DIFFERENT ANGLES SUPPLEMENTARY TO ∠AVB.

9. SHOW THAT ∠DCG AND ∠CGF ARE SUPPLEMENTARY.

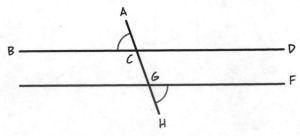

Chapter 6
THE TRIANGLE

Now we come to our headliner, the star of the show—and no, it isn't an actual star.

SAD!

The lead role in geometry has just **three sides**—but who ever said that celebrities had to be complicated?

Definition. IF A, B, AND C ARE THREE NONCOLLINEAR POINTS, THEN THE **TRIANGLE ABC**, OR **△ABC**, CONSISTS OF THE THREE LINE SEGMENTS AB, BC, AND AC. THE POINTS **A, B,** AND **C** ARE CALLED THE TRIANGLE'S **VERTICES,** AND THE SEGMENTS ARE ITS **SIDES.**

SIDE AB IS **ADJACENT** TO THE VERTEX A.

SIDE AC IS **ADJACENT** TO THE VERTEX A.

SIDE BC IS **OPPOSITE** VERTEX A.

WE BEGIN BY DOING SOMETHING THAT MAY SOUND ALMOST IDIOTICALLY SIMPLE: **COMPARING** TWO TRIANGLES TO SEE IF THEY HAVE THE SAME **SIZE** AND **SHAPE.**

GIVEN TRIANGLES △ABC AND △PQR, IMAGINE LAYING SOME CLEAR MATERIAL OVER △PQR.

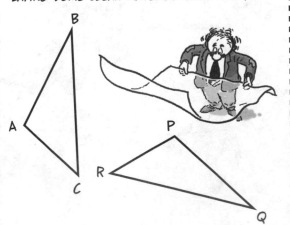

TRACE THE TRIANGLE △PQR ON THIS SHEET.

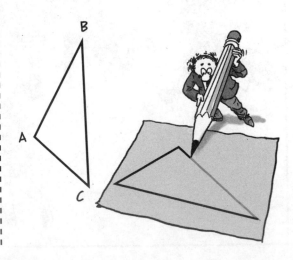

60

LIFT THE TRACING AND LAY IT OVER △ABC.

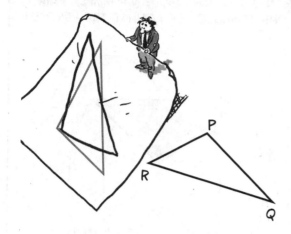

IF THE TRACING CAN BE MADE TO FIT PERFECTLY OVER △ABC, THEN △ABC AND △PQR HAVE THE SAME SIZE AND SHAPE.

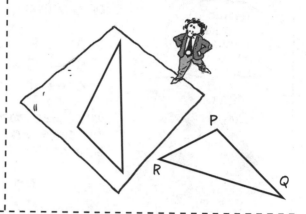

IT MAY BE THAT SIMPLY SLIDING THE TRACING ISN'T ENOUGH...

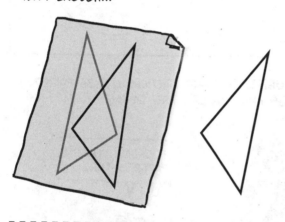

BUT FLIPPING THE SHEET OVER DOES CREATE A PERFECT MATCH.

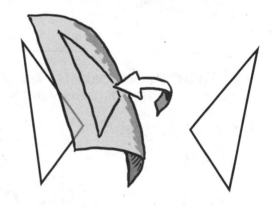

IF THIS **SUPERPOSITION,** WITH OR WITHOUT A FLIP, MAKES THE TRACING COINCIDE PERFECTLY WITH △ABC, THEN WE SAY THAT THE TRIANGLES ARE

CONGRUENT.

A WORD MUCH USED IN GEOMETRY BUT NOT SO MUCH OTHERWISE...

THESE THREE TRIANGLES ARE ALL CONGRUENT WITH EACH OTHER.

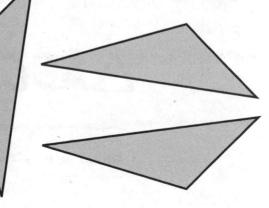

61

Congruence

WE NEED TO DESCRIBE CONGRUENCE IN LESS PHYSICAL TERMS THAN TRACING, LIFTING, ETC. HOW ABOUT WE SIMPLY **PAIR** THE VERTICES OF ONE TRIANGLE OFF AGAINST THE VERTICES OF ANOTHER?

IS THERE A SLIDE-AND-FLIP POSTULATE?

UM, NO...

A ↔ P
B ↔ Q
C ↔ R

WHICH ALSO PAIRS THE SIDES WITH EACH OTHER:

AB ↔ PQ
BC ↔ QR
AC ↔ PR

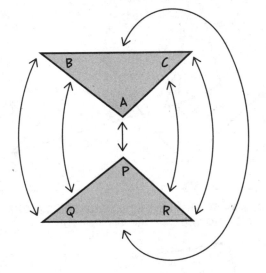

Definition. TWO TRIANGLES ARE **CONGRUENT** IFF THEIR VERTICES CAN BE PAIRED SO THAT ALL PAIRED (OR CORRESPONDING) **SIDES** AND **ANGLES** ARE EQUAL.

IN THE TRIANGLES ABOVE,

∠A = ∠P
∠B = ∠Q
∠C = ∠R
AB = PQ
BC = QR
AC = PR

SIX SEPARATE EQUALITIES!

TOO BAD... TRACING PAPER IS FUN...

THE SYMBOL FOR CONGRUENCE IS ≅, AS IN

$$\triangle ABC \cong \triangle PQR$$

A GLORIFIED EQUALS SIGN!

READ THIS AS "TRIANGLE A-B-C IS CONGRUENT TO TRIANGLE P-Q-R."

62

IT'S EASY TO SEE THAT IF TWO TRIANGLES ARE CONGRUENT TO A THIRD, THEN THEY'RE CONGRUENT TO EACH OTHER.

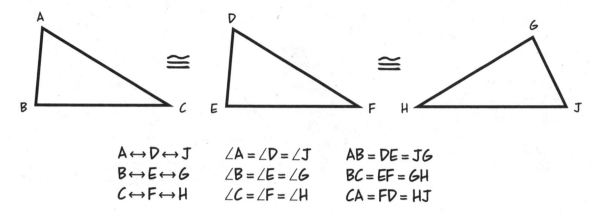

$$A \leftrightarrow D \leftrightarrow J \qquad \angle A = \angle D = \angle J \qquad AB = DE = JG$$
$$B \leftrightarrow E \leftrightarrow G \qquad \angle B = \angle E = \angle G \qquad BC = EF = GH$$
$$C \leftrightarrow F \leftrightarrow H \qquad \angle C = \angle F = \angle H \qquad CA = FD = HJ$$

CAUTION: ORDER MATTERS! WHEN WRITING A CONGRUENCE, THE SEQUENCE OF VERTICES ON ONE SIDE MUST MATCH THE CORRESPONDING VERTICES ON THE OTHER SIDE **IN ORDER.** IN THIS ILLUSTRATION,

$$\triangle ABC \cong \triangle PQR$$

$$NOT \ \triangle ABC \cong \triangle PRQ$$

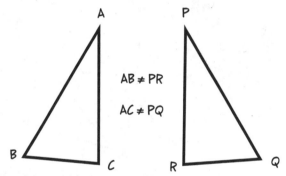

$$AB \neq PR$$
$$AC \neq PQ$$

BECAUSE THE CORRESPONDENCE INCLUDES NEITHER

$$B \leftrightarrow R \ NOR \ C \leftrightarrow Q$$

WE SHOULD BE ABLE TO READ THE CORRESPONDING PARTS DIRECTLY FROM THE CONGRUENCE. OTHERWISE, CONFUSION ENSUES!

GEOMETERS OF THE JURY, DO THE CORRESPONDENCES CORRESPOND? DOES THE CONGRUENCE LOOK INCON-GRUOUS? COULD ANY SANE PERSON CONCLUDE BEYOND A REASONABLE DOUBT ANY-THING BUT ETC. ETC. ETC.

ORDER! ORDER!

63

Congruence Tests

BY DEFINITION, **SIX** EQUALITIES MAKE TWO TRIANGLES CONGRUENT: ALL THREE CORRE- SPONDING SIDES AND ALL THREE CORRESPONDING ANGLES. IN PRACTICE, HOWEVER, ALL IT TAKES IS **THREE**—IF THEY'RE THE RIGHT THREE.

NOT THOSE THREE, I GUESS?

TWO TRIANGLES WITH THREE CORRESPONDING ANGLES THAT ARE EQUAL BUT NOT CONGRUENT.

SUPPOSE TWO TRIANGLES HAVE TWO PAIRS OF SIDES EQUAL, AS WELL AS THE ANGLE BETWEEN THEM. LET'S PLAY THE TRACING GAME WITH THESE TWO (SEE PP. 60–61).

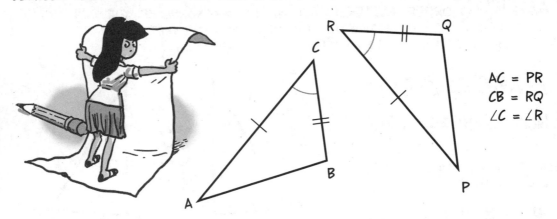

AC = PR
CB = RQ
∠C = ∠R

MOMO LAYS THE TRACING △A'B'C' OF △ABC OVER △PQR SO THAT A'C' LIES ON PR. BECAUSE ∠C = ∠R, C'B' MUST LIE EXACTLY ON RQ, AND BECAUSE THE SEGMENTS ARE EQUAL, B' MUST COINCIDE WITH Q. THE TRIANGLES APPEAR IDENTICAL.

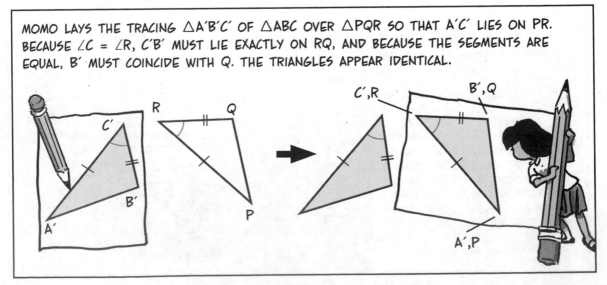

A VERY SIMILAR ARGUMENT WORKS WHEN THERE IS EQUALITY BETWEEN TWO PAIRS OF ANGLES AND THE SIDE BETWEEN THEM. SUPPOSE THAT:

$\angle A = \angle P$, $\angle C = \angle R$, $AC = PR$

THEN THE "LAYING ON OF TRIANGLES" PUTS A'C' OVER PR AND C'B' OVER RQ. THEIR INTERSECTION Q MUST LIE UNDER B'.

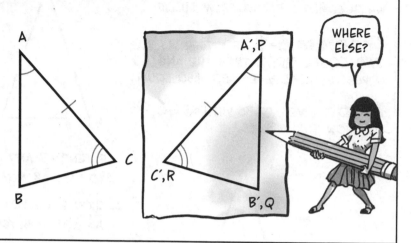

WHERE ELSE?

ALTHOUGH THESE ROUTINES MAY SATISFY OUR IMAGINATIONS, THEY HAVE NO LOGICAL BASIS—UNTIL WE TURN THEM INTO **NEW POSTULATES**.

Side-Angle-Side

Postulate 8 (SAS). IF TWO TRIANGLES HAVE TWO SIDES EQUAL AND THE ANGLES BETWEEN THEM EQUAL, THEN THE TRIANGLES ARE CONGRUENT.

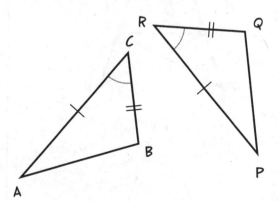

$$AC = PR, \; BC = QR, \; \angle C = \angle R$$
$$\Rightarrow$$
$$\triangle ABC \cong \triangle PQR$$

Angle-Side-Angle

Postulate 9 (ASA). IF TWO TRIANGLES HAVE TWO ANGLES AND THE SIDE BETWEEN THEM EQUAL, THEN THE TRIANGLES ARE CONGRUENT.

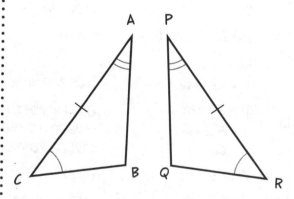

$$\angle A = \angle P, \; \angle C = \angle R, \; AC = PR$$
$$\Rightarrow$$
$$\triangle ABC \cong \triangle PQR$$

BOTH **SAS** AND **ASA** SHOW UP IN OUR
FIRST INDUSTRIAL-STRENGTH PROOF,
WHICH COMES STRAIGHT FROM EUCLID:

Theorem 6-1. IF A TRIANGLE HAS
TWO EQUAL SIDES, THEN THE **ANGLES**
OPPOSITE THOSE SIDES ARE ALSO EQUAL.

Proof. GIVEN △ABC WITH AB=AC,
WE SHOW THAT ∠B=∠C.

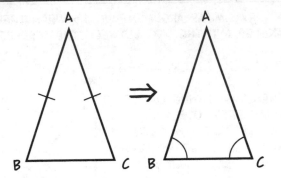

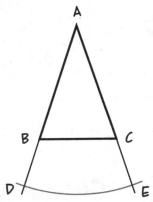

1. EXTEND AB AND AC   (RULER POST.)
   TO RAYS $\overrightarrow{AB}$ AND $\overrightarrow{AC}$.

2. CHOOSE ANY POINT D ON   (RULER POST.)
   $\overrightarrow{AB}$ WITH D BEYOND B.

3. WITH COMPASS CENTERED   (RULER POST.)
   AT A, FIND POINT E ON
   AC WITH AE=AD.

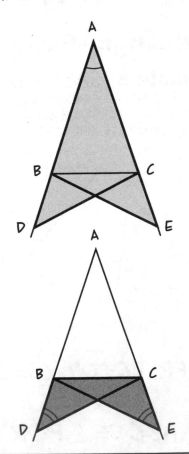

4. DRAW THE SEGMENTS   (POST. 2)
   BE AND CD.

5. ∠A = ∠A, AC = AB   (ASSUMED)

6. AD = AE   (STEP 3)

7. △**ABE** ≅ △**ACD**   (SAS)

8. THEN
   ∠D = ∠E AND CD = BE   (CORR. PARTS)

9. BD = AD – AB   (RULER POST.)
   = AE – AC   (SUBSTITUTION)
   = CE   (RULER POST.)

10. △**BCD** ≅ △**CBE**   (SAS)

11. ∠ECB = ∠DBC   (CORR. PARTS)

12. BUT THEN

   ∠**ABC** = 180° – ∠DBC   (PROTRACTOR POST.)
   = 180° – ∠ECB   (SUBSTITUTION)
   = ∠ACB ∎   (PROTRACTOR POST.)

66

THIS PROOF IS TYPICAL OF EUCLID: HE CLEVERLY DRAWS SOME EXTRA LINES, FINDS CONGRUENT TRIANGLES, AND USES THEIR CORRESPONDING PARTS TO PROVE SOME EQUALITY.

FIRST, THIS CONGRUENCE IMPLIED THESE EQUALITIES...

THEN THIS CONGRUENCE IMPLIED THIS EQUALITY.

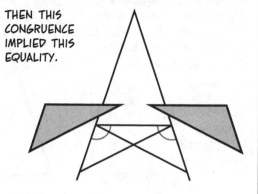

REMEMBER WHEN WE SAID TO BE ALERT TO THE POSSIBILITY OF TRUE **CONVERSES** (P. 23)? THEOREM 6-1 IS A GREAT EXAMPLE.

Theorem 6-2. IF A TRIANGLE HAS TWO EQUAL ANGLES, THEN IT HAS TWO EQUAL SIDES.

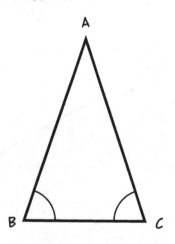

Proof. AGAIN WE START WITH △ABC, ONLY NOW WE ASSUME THAT ∠B = ∠C AND PROVE THAT AB = AC.

1. PAIR OFF THE VERTICES OF THE TRIANGLE WITH **THEMSELVES** LIKE THIS:

$$A \leftrightarrow A, \quad B \leftrightarrow C, \quad C \leftrightarrow B$$

2. ∠A = ∠A, ∠B = ∠C, BC = CB   (ASSUMED)

3. △**ABC** ≅ △**ACB** (!!!)   (ASA)

4. AB = AC ∎   (CORR. PARTS)

THIS PROOF AMOUNTS TO SAYING THAT A TRIANGLE WITH TWO EQUAL ANGLES IS CONGRUENT TO ITS OWN **MIRROR IMAGE** WITHOUT BEING FLIPPED OVER.

A TRIANGLE WITH TWO EQUAL SIDES IS CALLED **ISOSCELES.** THEOREMS 6-1 AND 6-2 SAY THAT **A TRIANGLE IS ISOSCELES IFF IT HAS TWO EQUAL ANGLES.**

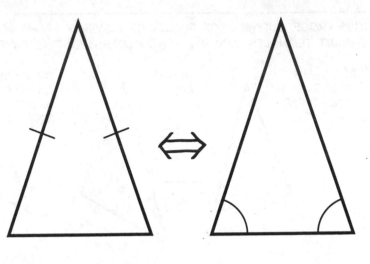

AND TWO EQUAL SCELES?

IN AN ISOSCELES TRIANGLE, THE SEGMENT BETWEEN THE EQUAL ANGLES IS THE **BASE;** THE TWO EQUAL SIDES ARE THE **LEGS;** THE VERTEX WHERE THEY MEET IS THE **APEX.**

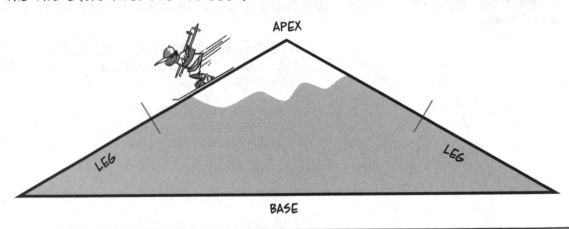

APEX

LEG

LEG

BASE

IF AN ISOSCELES TRIANGLE IS SPLIT DOWN THE MIDDLE, ONE HALF LOOKS LIKE THE MIRROR IMAGE OF THE OTHER HALF. WHERE IS THAT MIRROR, THAT CENTRAL LINE?

IF THOSE TWO AREN'T CONGRUENT, I'LL EAT MY HAT!

68

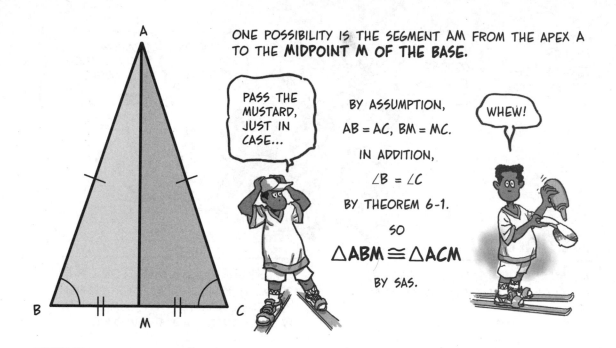

ONE POSSIBILITY IS THE SEGMENT AM FROM THE APEX A TO THE **MIDPOINT M OF THE BASE.**

PASS THE MUSTARD, JUST IN CASE...

BY ASSUMPTION, AB = AC, BM = MC.

IN ADDITION,

∠B = ∠C

BY THEOREM 6-1.

SO

△ABM ≅ △ACM

BY SAS.

WHEW!

THE TWO HALVES OF AN ISOSCELES TRIANGLE ARE IN FACT CONGRUENT, SO ALL THEIR CORRESPONDING PARTS MUST BE EQUAL:

∠BAM = ∠CAM

∠AMB = ∠AMC

THE FIRST EQUATION SAYS THE APEX ANGLE IS SPLIT EQUALLY; THE SECOND SAYS THAT ∠AMB AND ∠AMC, A LINEAR PAIR, MUST BOTH BE 180°/2 = **90°.** THIS PROVES:

Theorem 6-3. IN AN ISOSCELES TRIANGLE, THE LINE JOINING THE APEX TO THE MIDPOINT OF THE BASE IS **PERPENDICULAR** TO THE BASE AND DIVIDES THE APEX ANGLE IN **HALF.** ∎

A VERY SPECIAL LINE...

NICE...

Definitions

AN **ANGLE BISECTOR** IS A LINE, RAY, OR SEGMENT THAT DIVIDES AN ANGLE IN HALF.

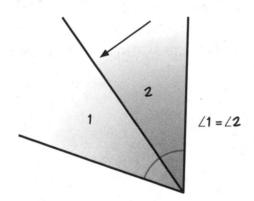

∠1 = ∠2

A **PERPENDICULAR BISECTOR** OF A SEGMENT IS A LINE, RAY, OR SEGMENT PERPENDICULAR TO THE SEGMENT AND PASSING THROUGH ITS MIDPOINT.

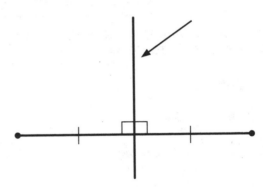

IN A TRIANGLE, A **MEDIAN** IS A SEGMENT, RAY, OR LINE JOINING ANY VERTEX WITH THE MIDPOINT OF THE OPPOSITE SIDE.

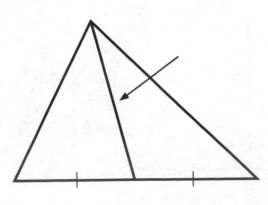

IN THESE TERMS, THEOREM 6-3 SAYS: IN AN **ISOSCELES** TRIANGLE, THE **MEDIAN** FROM THE APEX, THE **ANGLE BISECTOR** AT THE APEX, AND THE **PERPENDICULAR BISECTOR** OF THE BASE ARE ALL THE **SAME LINE.**

A SORT OF CONVERSE IS ALSO TRUE.

Theorem 6-4. IN ANY TRIANGLE, IF A PERPENDICULAR BISECTOR OF ANY SIDE IS ALSO A MEDIAN, THEN THE TRIANGLE IS ISOSCELES.

Proof. IN △ABC, LET M BE A POINT ON BC WITH BM=MC AND ∠AMB=∠AMC=90°. WE SHOW THAT AB=AC.

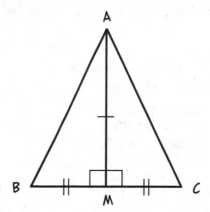

1. AM = AM, BM = MC,   (ASSUMED)
   ∠AMB = ∠AMC

2. △ABM ≅ △ACM   (SAS)

3. AB = AC ▮   (CORR. PARTS)

AB≠AC IFF THIS HAPPENS...

NICE. VERY NICE.

Corollary 6-4.1. ANY POINT ON A SEGMENT'S PERPENDICULAR BISECTOR IS EQUIDISTANT FROM THE SEGMENT'S ENDPOINTS.

Proof. IN THE DIAGRAM, WE ASSUME AM=MB AND PM⊥AB, AND PROVE PA=PB.

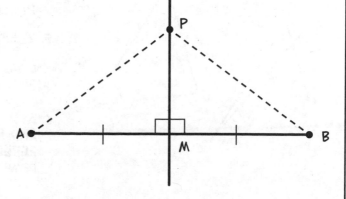

1. PM IS A MEDIAN   (DEFINITION)
   OF △APB.

2. PM IS THE   (ASSUMED)
   PERPENDICULAR
   BISECTOR OF AB.

3. △APB IS   (THM. 6-4)
   ISOSCELES,
   WITH PA=PB. ▮

71

Side-Side-Side

Theorem 6-5 (SSS). IF TWO TRIANGLES HAVE **ALL THREE PAIRS OF CORRESPONDING SIDES** EQUAL, THEN THE TRIANGLES ARE CONGRUENT.

Proof. GIVEN △ABC AND △DEF WITH AB=DE, AC=DF, BC=EF, WE PROVE △ABC ≅ △DEF.

 THIS WILL BE OUR FIRST **INDIRECT PROOF**, OR **PROOF BY CONTRADICTION** (SEE P. 25). IN STEP 6, WE MAKE AN ASSUMPTION THAT LEADS TO A CONTRADICTION, SO WE CONCLUDE THAT THE ASSUMPTION MUST BE FALSE.

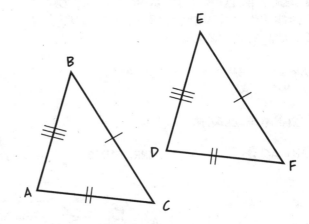

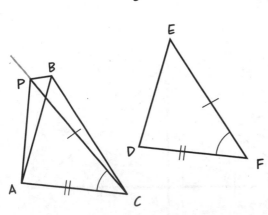

FIRST WE COPY THE ANGLE ∠F ON THE SEGMENT AC.

1. THERE IS A LINE $\overline{CX}$ SUCH THAT ∠XCA=∠F IN THE SAME HALF-PLANE OF AC AS POINT B.   (PROTR. POST.)

2. THERE IS A POINT **P** ON $\overline{CX}$ WITH PC=EF.   (RULER POST.)

3. DRAW PA.   (POST. 1)

4. AC=DF   (ASSUMED)
   PC=EF, ∠PCA=∠F   (STEPS 1, 2)

5. △**APC** ≅ △**DEF**   (SAS)

NOW WE MAKE THE FATEFUL ASSUMPTION:

6. **ASSUME THAT P AND B ARE DIFFERENT POINTS.**   (FOR THE SAKE OF ARGUMENT!)
   DRAW SEGMENT PB.   (POST. 1)

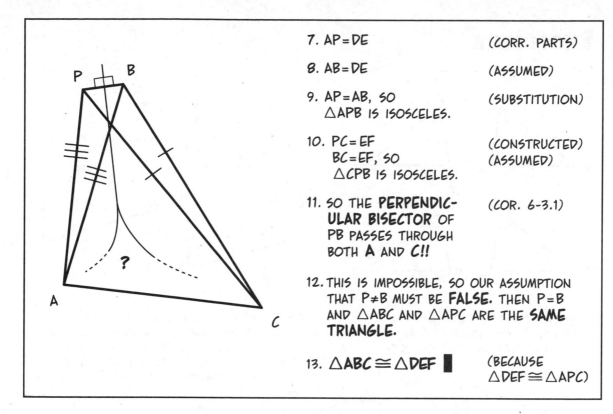

7. AP = DE   (CORR. PARTS)

8. AB = DE   (ASSUMED)

9. AP = AB, SO   (SUBSTITUTION)
   △APB IS ISOSCELES.

10. PC = EF   (CONSTRUCTED)
    BC = EF, SO   (ASSUMED)
    △CPB IS ISOSCELES.

11. SO THE **PERPENDIC-   (COR. 6-3.1)
    ULAR BISECTOR** OF
    PB PASSES THROUGH
    BOTH **A** AND **C**!!

12. THIS IS IMPOSSIBLE, SO OUR ASSUMPTION
    THAT P≠B MUST BE **FALSE.** THEN P=B
    AND △ABC AND △APC ARE THE **SAME
    TRIANGLE.**

13. △**ABC** ≅ △**DEF** ▌   (BECAUSE
                                      △DEF ≅ △APC)

SIDE-SIDE-SIDE IS THE THIRD OF THIS CHAPTER'S CONGRUENCE TESTS (WITH MORE TO
COME IN FUTURE CHAPTERS). ALONG WITH SAS AND ASA, IT SHOWS HOW TO PROVE
CONGRUENCE USING JUST THREE EQUALITIES RATHER THAN SIX.

WE CAN CRACK PROBLEMS IN HALF THE STEPS!

NICE.

WHAT EVER HAPPENED TO THE VALUE OF OVER-WORK?

IN MY DAY, WE USED TO DO GEOMETRY WITH SCORPIONS IN OUR UNDERWEAR...

CONGRUENCE IS A POWERFUL TOOL.
THEOREMS 6-1 THROUGH 6-4 USED IT TO SHOW EQUALITIES WITHIN GEOMETRICAL FIGURES. IN THE NEXT CHAPTER, WE USE CONGRUENCE TO PROVE SOME IMPORTANT AND USEFUL **INEQUALITIES,** TOO.

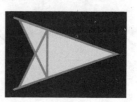

Exercises

1. IS △ABC ≅ △PQR?

a.

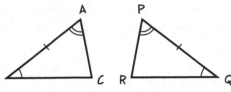

b.

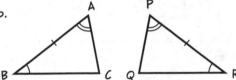

c.

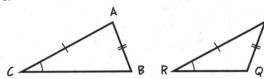

d.

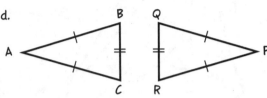

2. IN THIS DIAGRAM, THE SEGMENTS AP AND BQ INTERSECT AT C. IS △ABC ≅ PQC? WHY OR WHY NOT?

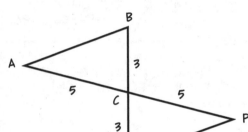

3. THERE'S A REASON WE NEVER MENTIONED CONGRUENCE AND **SSA**, WHERE THE ANGLE ISN'T BETWEEN THE SIDES. CAN YOU FINISH THE TRIANGLE △PQR WITH ∠B=∠Q, AB=PQ, AND AC=PR, BUT △PQR IS **NOT** CONGRUENT TO △ABC? (HINT: USE A COMPASS.)

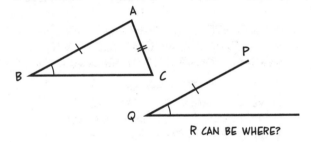

R CAN BE WHERE?

4. SUPPOSE POINTS B AND C ARE EQUIDISTANT FROM A, AS SHOWN. DRAW BC, MAKING △ABC. IS A ON THE PERPENDICULAR BISECTOR OF BC? WHY OR WHY NOT?

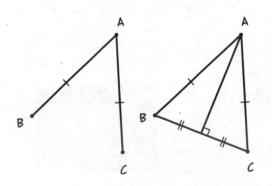

5. GIVEN A CORRESPONDENCE BETWEEN △ABC AND △PQR, AND ANOTHER CORRESPONDENCE BETWEEN △PQR AND △TUV, WHAT'S A "NATURAL" CORRESPONDENCE BETWEEN △ABC AND △TUV?

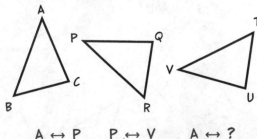

A ↔ P	P ↔ V	A ↔ ?
B ↔ R	Q ↔ U	B ↔ ?
C ↔ Q	R ↔ T	C ↔ ?

Chapter 7
INEQUALITIES IN TRIANGLES

SOME THINGS BEING NOT SO EQUAL...

TWO QUANTITIES, WHETHER THEY'RE SEGMENTS, ANGLES, OR THE MASSES OF ELEPHANTS, ARE USUALLY **NOT** EQUAL.

THEN WHY SO MUCH TALK ABOUT EQUALITY IN MATH?

UM... 'CAUSE IT'S RARE?

A FEW THINGS ABOUT INEQUALITY TO REMEMBER FROM ARITHMETIC:

IF $a>b$ AND $b>c$, THEN $a>c$. IT'S OKAY TO WRITE $a>b>c$.

IF $a>b$ AND $c \geq d$, THEN $a+c>b+d$. AN ELEPHANT CARRYING A PIANO WEIGHS MORE THAN A MOUSE LIFTING A PEA (OR A PIANO).

A PART IS LESS THAN THE WHOLE. AN ELEPHANT WEIGHS MORE THAN ITS TUSK.

OUR FIRST INEQUALITY INVOLVES
ANGLES OUTSIDE OF TRIANGLES.

Definition. AN **EXTERIOR ANGLE** OF A TRIANGLE IS AN ANGLE THAT FORMS A
LINEAR PAIR WITH AN INTERIOR ANGLE. THE TWO OTHER INTERIOR ANGLES ARE CALLED
REMOTE FROM THE EXTERIOR ANGLE.

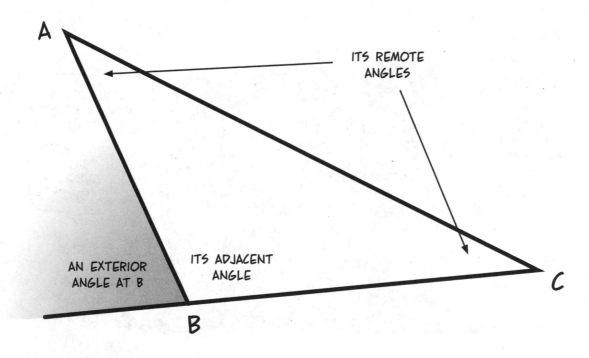

ITS REMOTE
ANGLES

AN EXTERIOR
ANGLE AT B

ITS ADJACENT
ANGLE

AN EXTERIOR ANGLE MAY BE LARGER OR SMALLER THAN ITS ADJACENT INTERIOR ANGLE.
HERE, FOR INSTANCE:

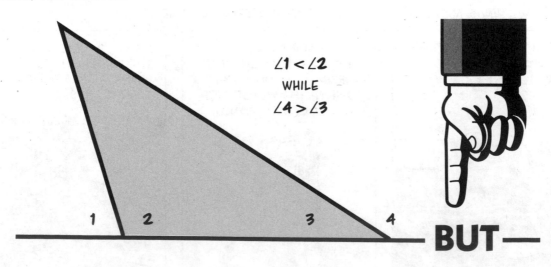

$\angle 1 < \angle 2$
WHILE
$\angle 4 > \angle 3$

BUT—

Theorem 7-1. AN EXTERIOR ANGLE OF A TRIANGLE IS GREATER THAN EITHER **REMOTE** INTERIOR ANGLE.

Proof. GIVEN △ABC WITH EXTERIOR ANGLE ∠ACD, WE SHOW THAT ∠ACD > ∠A, ∠ACD > ∠B. THE IDEA IS TO MAKE A CONGRUENCE WITH AN ANGLE THAT'S **PART** OF ∠ACD.

1. LET M BE THE MIDPOINT OF AC.   (RULER POST.)

2. DRAW RAY $\overrightarrow{BM}$ AND MARK E WITH ME = BM.   (RULER POST., THM. 4-1)

3. DRAW CE.   (POST. 2)

4. ∠AMB = ∠CME   (VERT. ∠s)

5. △AMB ≅ △CME   (SAS)

6. ∠A = ∠MCE   (CORR. PARTS)

7. ∠ACD > ∠MCE   (WHOLE > PART)

8. ∠ACD > ∠A   (SUBSTITUTION)

9. THE SAME CONSTRUCTION USING THE MIDPOINT OF BC SHOWS THAT ∠ACD > ∠B. ∎

THIS IS ANOTHER PROOF STRAIGHT FROM EUCLID.

MORE EXTRA LINES!

MORE COPIED SEGMENTS!

MORE CONGRUENT TRIANGLES!

EASY, BIG FELLA!

77

THEOREM 7-1 PAYS AN
IMMEDIATE DIVIDEND.

Theorem 7-2. IN A TRIANGLE, LONGER SIDES ARE OPPOSITE LARGER ANGLES.
IN △ABC, **AC>AB** ⇔ ∠B>∠C.

Proof. WE FIRST ASSUME
AC>AB AND PROVE THAT ∠B >∠C.

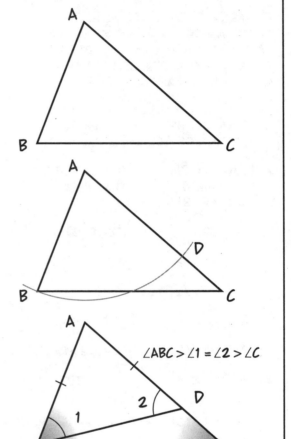

1. AC > AB   (ASSUMED)

2. MARK D ON AC   (THM. 4-1)
   WITH AD = AB.

3. DRAW BD.   (POST. 2)

4. △ABD IS ISOSCELES.   (DEFINITION)

5. ∠1 = ∠2   (THM. 6-1)

6. ∠1 < ∠ABC   (PART<WHOLE)

7. ∠2 < ∠ABC   (SUBSTITUTION)

8. ∠2 IS AN EXTERIOR   (DEF. OF EXT. ∠)
   ANGLE OF △BDC.

9. ∠C < ∠2   (THM. 7-1)

10. ∠C < ∠ABC ▌   (ARITHMETIC)

THE CONVERSE IS LEFT AS AN EXERCISE.

∠ABC > ∠1 = ∠2 > ∠C

IT MAKES PERFECT SENSE TO ME! THE BIGGER THE SPREAD, THE LONGER THE SIDE! AMIRIGHT?

WELL, SORT OF, YES, BUT IT'S NOT QUITE THAT SIMPLE...

MORE ON THIS IN THE EXERCISES.

78

Distance from Point to Line

LET'S APPLY THEOREMS 7-1 AND 7-2 TO **RIGHT** TRIANGLES, WHERE ONE ANGLE IS 90°.

IN A RIGHT TRIANGLE, THE OTHER TWO ANGLES ARE **ACUTE** (<90°). THAT'S BECAUSE AN **EXTERNAL** ANGLE AT THE RIGHT ANGLE IS ALSO RIGHT. BY THEOREM 7-1, THE TWO REMOTE ANGLES ARE SMALLER.

$$\angle A < 90°$$
$$\angle B < 90°$$

AND SO:

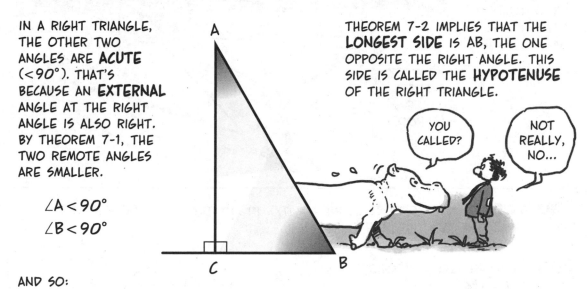

THEOREM 7-2 IMPLIES THAT THE **LONGEST SIDE** IS AB, THE ONE OPPOSITE THE RIGHT ANGLE. THIS SIDE IS CALLED THE **HYPOTENUSE** OF THE RIGHT TRIANGLE.

YOU CALLED?

NOT REALLY, NO...

Theorem 7-3. IF $\overline{AB}$ IS A LINE, AND C IS A POINT NOT ON $\overline{AB}$, AND $\angle CAB = 90°$, THEN **CA < CB**.

IN OTHER WORDS, THE **PERPENDICULAR** SEGMENT FROM C TO THE LINE IS THE **SHORTEST** SEGMENT FROM C TO THE LINE.

Proof. △ABC IS A RIGHT TRIANGLE WITH $\angle CAB = 90°$, SO CB, THE SIDE OPPOSITE IT, IS THE LONGEST SIDE. ∎

THE LENGTH OF THIS PERPENDICULAR SEGMENT IS CALLED THE **DISTANCE** FROM THE POINT TO THE LINE.

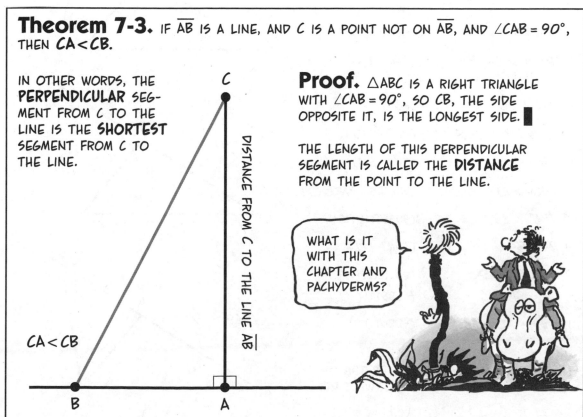

WHAT IS IT WITH THIS CHAPTER AND PACHYDERMS?

CA < CB

DISTANCE FROM C TO THE LINE $\overline{AB}$

THE FINAL
INEQUALITY IS
TOTALLY OBVIOUS.
IT SAYS THAT A
STRAIGHT LINE IS
ALWAYS LONGER
THAN A DETOUR.

BUT IT NEEDS PROOF!

Theorem 7-4 (THE TRIANGLE INEQUALITY). IN ANY TRIANGLE, THE SUM OF ANY TWO SIDES IS GREATER THAN THE THIRD.

Proof. GIVEN △ABC, WE SHOW THAT BA + AC > BC.

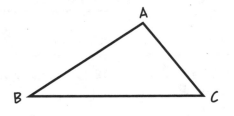

1. DRAW RAY $\overrightarrow{BA}$. (RULER POST.)

2. MARK D ON $\overrightarrow{BA}$ WITH AD = AC. (THM. 4-1)

3. DRAW DC. (POST. 2)

4. △ADC IS ISOSCELES. (DEFINITION)

5. ∠D = ∠1 (THM. 6-1)

6. ∠BCD > ∠1 (WHOLE > PART)

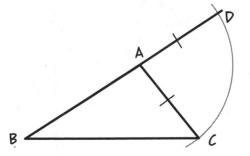

7. ∠BCD > ∠D (SUBSTITUTION)

8. BD > BC (THM. 7-2)

9. BD = BA + AD (RULER POST.)
   = BA + AC (SUBST.)

10. **BA + AC > BC** ▌ (SUBST.)

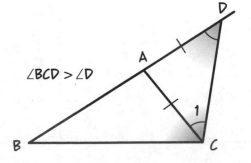

∠BCD > ∠D

11. A SIMILAR ARGUMENT PROVES THE IN-
    EQUALITY FOR THE OTHER TWO SIDES.

NOT LONG AFTER EUCLID PUBLISHED THIS PROOF, SOME WAGGISH GEOMETER CRACKED THAT EVEN A DONKEY KNEW ENOUGH TO GO STRAIGHT TO ITS FOOD.

UNLESS THERE'S ANOTHER ATTRACTION AT POINT B!

A HUMAN, APPARENTLY, REQUIRES A TEN-STEP PROOF.

WITHOUT A PROOF, WE MIGHT BE TEMPTED TO IMAGINE SOME FUNHOUSE GEOMETRY WHERE LINES BEND SO MUCH THAT ANYTHING COULD HAPPEN.

LUCKILY, THE THEOREM IS TRUE, AND DONKEYS CAN EAT IN PEACE.

WHAT A RELIEF!

STILL, DON'T RELAX TOO MUCH! EXOTIC GEOMETRIES DO EXIST, AND THEY'RE GOING TO FORCE ANOTHER POSTULATE ON US, JUST TO KEEP US OUT OF BIZARRO-WORLD.

DOES THIS GEOMETRY MAKE ME LOOK FLAT?

BUT FIRST, SOMETHING COMPLETELY DIFFERENT!

Exercises

1. THERE ARE ACTUALLY TWO EXTERIOR ANGLES AT EACH VERTEX OF A TRIANGLE. WHY ARE THEY EQUAL?

2. WHAT'S THE VALUE OF THE VERTICAL ANGLE(S) SHOWN HERE?

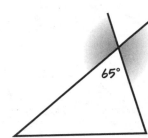

3a. IN THIS TRIANGLE, HOW BIG CAN ∠A BE? THAT IS, WHAT VALUE MUST IT NOT EXCEED?

3b. WHAT'S AN UPPER LIMIT ON THE SUM OF THE (INTERIOR) ANGLES OF THIS TRIANGLE?

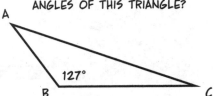

4. CONVERSE OF THEOREM 7-2. ASSUMING ∠B>∠C, LET'S PROVE THAT AC>AB. SUPPOSE THE CONTRARY IS TRUE, THAT AC **ISN'T** GREATER THAN AB. THEN THERE ARE TWO POSSIBILITIES: EITHER AB=AC OR AB>AC.

a. IF AB=AC, THEN... ?

b. IF AB>AC, THEN... ?

EITHER CASE CONTRADICTS THE HYPOTHESIS, SO WE CONCLUDE THAT AC>AB.

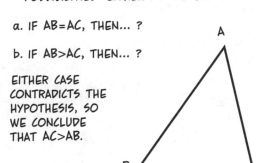

5. MOMO IS LOOKING FROM POINT O AT A VERTICAL POST. POINT C IS AT HER EYE LEVEL. TO PROVE: IF ∠AOB=∠BOC, THEN AB>BC.

a. OA>OC. WHY?

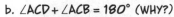

b. TAKE D ON OA WITH OD=OC. WHAT TRIANGLES ARE CONGRUENT, AND WHY?

c. WHY IS AB>BD? WHY IS AB>BC?

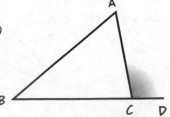

6. ACCORDING TO THEOREM 7-4, WHICH OF THESE TRIPLES MIGHT BE THE SIDES OF A TRIANGLE? WHICH ONES CAN'T?

a. (93, 58, 109)   b. (6, 6, 11)

c. (5, 2, 9)   d. (0.1, 0.2, 0.3)

e. (1,000,523, 1,000,525, 3)

f. IF ONE SIDE OF AN ISOSCELES TRIANGLE IS 8, WHAT MUST BE TRUE OF THE OTHER TWO SIDES?

7. LET'S SHOW THAT ANY PAIR OF ANGLES IN A TRIANGLE ADDS TO LESS THAN 180°.

a. EXTEND BC TO MAKE AN EXTERIOR ANGLE. THEN

∠ACD>∠B (WHY?) OR

∠B−∠ACD<0

b. ∠ACD + ∠ACB = 180° (WHY?)

c. ∠ACB + ∠B < 180° (ADDING)

d. FROM THIS, SHOW THAT ∠A+∠B+∠C < 270°.

Chapter 8
CLASSICAL CONSTRUCTIONS

Let's give our laboring brains a rest for a while and do a little work with our hands. In this chapter, we learn how to construct geometrical diagrams with a straightedge and compass.

Copying an Angle

GIVEN AN ANGLE ∠O AND A POINT P, WE WILL MAKE AN EQUAL ANGLE WITH VERTEX P.

DRAW ANY LINE L THROUGH P.

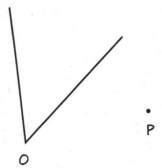

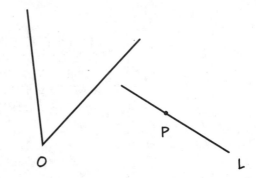

CENTERING COMPASS AT O, MARK A AND B ON THE RAYS OF ∠O WITH OA=OB. DRAW AN ARC OF THE SAME RADIUS CENTERED AT P, INTERSECTING L AT Q, SO PQ=OA.

CENTER COMPASS AT Q; DRAW ARC OF RADIUS AB INTERSECTING FIRST ARC AT R. DRAW PR. THEN ∠RPQ=∠AOB.

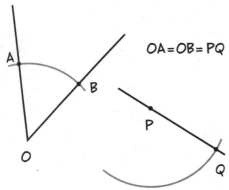

OA=OB=PQ

AB = QR
PR = PQ

WHY ARE THE ANGLES EQUAL? BY CONSTRUCTION,

OA=PR, OB=PQ, AB=RQ

SO

△AOB ≅ △RPQ BY SSS.

THEN ∠O = ∠P AS CORRESPONDING PARTS OF CONGRUENT TRIANGLES.

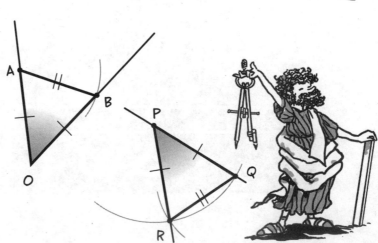

84

Bisecting an Angle

STARTING WITH AN ANGLE ∠V, CENTER THE COMPASS AT V AND DRAW AN ARC INTERSECTING ONE RAY AT A AND THE OTHER AT B.

DRAW ARCS OF EQUAL RADIUS CENTERED AT A AND B, INTERSECTING AT C.

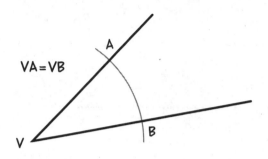

VA=VB

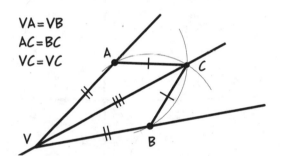

AC=BC

DRAW $\overline{VC}$. THIS IS THE ANGLE BISECTOR.

WHY? BECAUSE, AS BEFORE, △VAC ≅ △VBC BY SSS, SO ∠AVC=∠BVC AS CORR. PARTS.

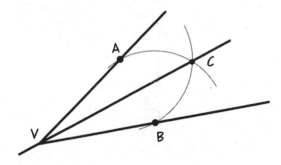

VA=VB
AC=BC
VC=VC

Perpendicular Bisector of a Segment

GIVEN SEGMENT AB, DRAW TWO ARCS, CENTERED AT A AND B, OF EQUAL RADIUS LARGE ENOUGH TO INTERSECT AT C AND D. THE LINE $\overline{CD}$ IS THE PERPENDICULAR BISECTOR.

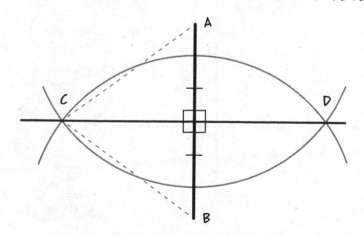

WHY? BECAUSE, $\overline{CD}$ BISECTS THE ANGLE ∠ACB, PER THE PREVIOUS CONSTRUCTION. BUT ALSO △CAB IS ISOSCELES, SO THE ANGLE BISECTOR AT ITS APEX IS THE PERPENDICULAR BISECTOR OF THE BASE (THEOREM 6-3).

Perpendicular Through a Point

GIVEN A LINE L AND A POINT P, WE DRAW A NEW LINE M CONTAINING P WITH M⊥L.

LOOKS LIKE A GOOD PLACE TO START...

CENTERING COMPASS AT P, DRAW AN ARC THAT INTERSECTS THE LINE AT A AND B.

P •

PA = PB

DRAW ARCS OF EQUAL RADIUS CENTERED AT A AND B AND INTERSECTING AT Q. THEN PQ⊥AB.

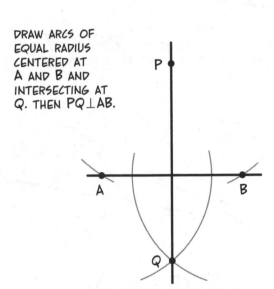

AS BEFORE, PQ BISECTS THE ANGLE ∠APB.

△APB IS ISOSCELES, SO THIS ANGLE BISECTOR IS PERPENDICULAR TO THE BASE AB.

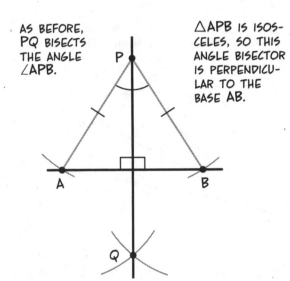

THIS CONSTRUCTION WORKS JUST AS WELL IF P IS ON THE LINE AS OFF IT.

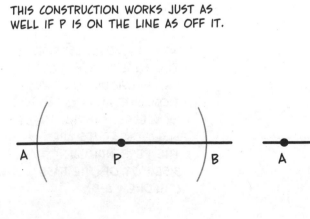

PERFECTLY PERPETRATED PERPENDICULARS!

Circle Through Three Noncollinear Points

ANY THREE NONCOLLINEAR POINTS P, Q, AND R ARE THE VERTICES OF A TRIANGLE.

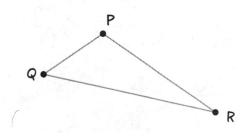

CONSTRUCT THE PERPENDICULAR BISECTORS OF TWO SIDES OF △PQR, SAY PQ AND PR. SUPPOSE THEY INTERSECT AT C.

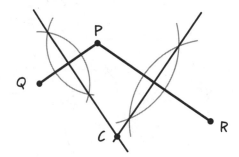

POINTS ON THE PERPENDICULAR BISECTOR OF A SEGMENT ARE EQUIDISTANT FROM ITS END-POINTS, BY COR. 6-4.1.

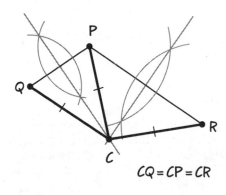

$$CQ = CP = CR$$

SO THE CIRCLE CENTERED AT C WITH RADIUS CP ALSO PASSES THROUGH Q AND R.

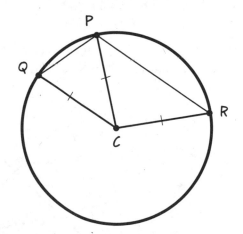

THIS IS THE **ONLY** CIRCLE CONTAINING P, Q, AND R. WHY? IF SOME OTHER CIRCLE CONTAINED THE POINTS, ITS CENTER C' WOULD BE EQUIDISTANT FROM P, Q, AND R. SO C' LIES ON THE PERPEN-DICULAR BISECTORS OF PQ AND PR.* THAT IS, IT'S AT THEIR INTER-SECTION, AND C' = C.

Three noncollinear points determine a circle.

DETERMINED TO DO **WHAT?**

*BY PROBLEM 4 IN THE EXERCISES TO CHAPTER 6 ON PAGE 74.

IT DOESN'T MATTER WHICH TWO SIDES OF △PQR ARE USED IN THAT CONSTRUCTION. THE PERPENDICULAR BISECTORS OF ANY PAIR OF SIDES MUST INTERSECT AT THE **SAME CENTER C.** ONLY ONE CIRCLE CAN PASS THROUGH ALL THREE POINTS.

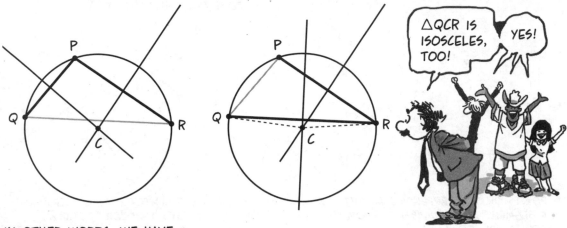

△QCR IS ISOSCELES, TOO!

YES!

IN OTHER WORDS, WE HAVE FOUND THIS SURPRISING RESULT:

Theorem 8-1. THE PERPENDICULAR BISECTORS OF THE THREE SIDES OF **ANY TRIANGLE** MEET AT A **SINGLE POINT.**

EEHAH! THAT IS SO **COOL** AND TOTALLY **UNEXPECTED!** PASS THE MUSTARD! TIME TO **CELEBRATE!**

UM... NOT QUITE YET, SORRY...

THE HAT AGAIN? REALLY?

UNFORTUNATELY, IT TURNS OUT WE HAVE A LITTLE **PROBLEM...**

LET'S LOOK CLOSELY AT PAGE 87, THE CONSTRUCTION OF A CIRCLE THROUGH THREE POINTS.

THE FIRST STEP DRAWS ARCS OF EQUAL RADIUS CENTERED AT TWO OF THE POINTS.

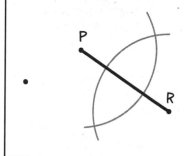

POSTULATE 6 (P. 43) ENSURES THAT THESE ARCS INTERSECT IN TWO POINTS, WHEN THE RADII ARE BIG ENOUGH.

SO FAR, SO GOOD!

BY POSTULATE 1, WE CAN JOIN THESE POINTS. THE RESULT IS A PERPENDICULAR BISECTOR.

THEN WE DID THE SAME FOR A SECOND PAIR OF POINTS.

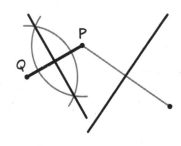

FINALLY, WE USED THE POINT WHERE THE TWO PERPENDICULAR BISECTORS INTERSECT AS THE CIRCLE'S CENTER.

C →

AND THE QUESTION IS...

How do we know that they intersect?

OOPS! ER...

BECAUSE? IS "BECAUSE" GOOD ENOUGH?

I'M HAVING A HOT DOG ANYWAY!

THIS LITTLE DETAIL, IT TURNS OUT, REVEALS A **BIG GAP** IN OUR POSTULATES, A GAP THAT WILL REQUIRE THE NEXT **TWO CHAPTERS** TO FILL.

OH, ONLY A HOT DOG...

Exercises

TIME TO PLAY WITH STRAIGHTEDGE AND COMPASS! EXPECT TO USE SEVERAL SHEETS OF PAPER, BECAUSE IT'S EASIER TO WORK AT FAIRLY LARGE SCALE.

1. DRAW SOME LINE SEGMENTS AND CONSTRUCT THEIR PERPENDICULAR BISECTORS.

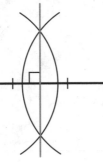

2. DRAW SOME ANGLES AND CONSTRUCT THEIR ANGLE BISECTORS.

3. GIVEN SEGMENT PQ, (a) FIND ITS MIDPOINT. (b) CONSTRUCT A PERPENDICULAR AT Q. (c) CONSTRUCT SEGMENT QR WITH QR=$\frac{1}{2}$PQ.

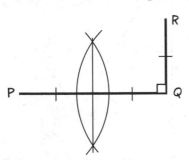

4. (EUCLID'S FIRST CONSTRUCTION.) GIVEN SEGMENT AB, DRAW ARCS OF LENGTH AB CENTERED AT EACH ENDPOINT. JOIN EACH ENDPOINT TO THE ARCS' INTERSECTION C.

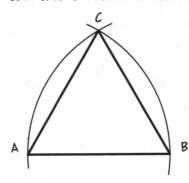

WHY IS △ABC **EQUILATERAL**, THAT IS, ALL ITS SIDES ARE EQUAL?

5. GIVEN LINE $\overline{AB}$ AND POINT P NOT ON L, DRAW $\overline{AP}$. NOW COPY ∠PAB AT P, WITH ONE SIDE ON $\overline{AP}$. COMPARING THE ANGLE'S OTHER SIDE WITH $\overline{AB}$, DO YOU SEE ANYTHING SPECIAL ABOUT THEM?

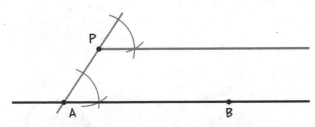

6. GIVEN SEGMENTS AB, CD > $\frac{1}{2}$AB, CONSTRUCT AN ISOSCELES TRIANGLE WITH BASE AB AND LEG = CD.

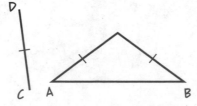

7. IN AN EQUILATERAL TRIANGLE, DRAW PERPENDICULAR BISECTORS OF ALL SIDES. THEY MEET IN A POINT M (WHY?) AND PASS THROUGH THE VERTICES (WHY?).

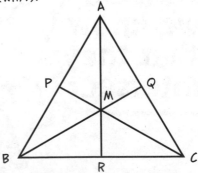

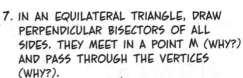

SHOW THAT △AQB ≅ △CPB.
SHOW THAT △BRM ≅ △AQM.
DO YOU SEE ANY OTHER CONGRUENCES?

Chapter 9
THE INTERSECTION PROBLEM
A BENT CHAPTER, BUT NOT A POLITICAL ONE

SOMETIMES A BUILDER WANTS
TO MAKE LINES THAT INTERSECT,
AND SOMETIMES NOT.

WHEN DO TWO LINES INTERSECT? AND WHEN DO THEY NOT?

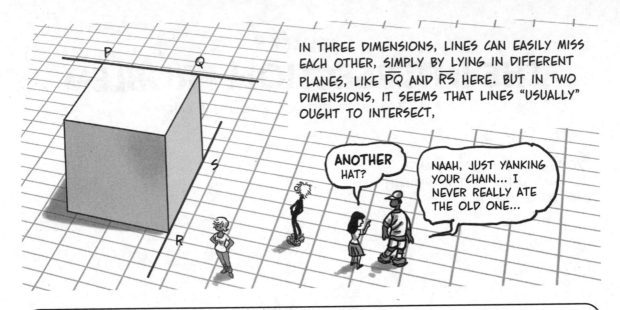

IN THREE DIMENSIONS, LINES CAN EASILY MISS EACH OTHER, SIMPLY BY LYING IN DIFFERENT PLANES, LIKE $\overline{PQ}$ AND $\overline{RS}$ HERE. BUT IN TWO DIMENSIONS, IT SEEMS THAT LINES "USUALLY" OUGHT TO INTERSECT,

ANOTHER HAT?

NAAH, JUST YANKING YOUR CHAIN... I NEVER REALLY ATE THE OLD ONE...

GIVEN NONCOLLINEAR POINTS A, B, AND C, WE CAN ALWAYS DRAW INTERSECTING LINES BY CONNECTING TWO DIFFERENT PAIRS OF POINTS, BY POSTULATE 1.

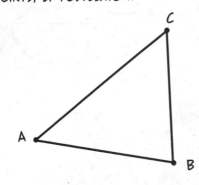

BUT WHAT ABOUT SOME OTHER LINE M THROUGH C IN THE SAME PLANE AS A AND B? HOW CAN WE TELL WHETHER M INTERSECTS $\overline{AB}$?

ER... I THINK IT DOES...

THE INTERSECTION, IF IT EXISTS, MIGHT BE A BILLION TRILLION MILES AWAY. HOW ARE WE SUPPOSED TO KNOW?

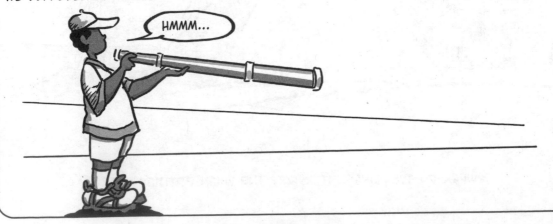

HMMM...

SOMEHOW, ASKING ABOUT INTERSECTIONS RAISES QUESTIONS ABOUT DISTANT REGIONS OF **SPACE ITSELF.** WHAT'S IT LIKE OUT THERE ON THE WAY TO INFINITY?

AND WHO HAS TIME TO FIND OUT?

CLOSER TO HOME, WE CAN TRY SOMETHING REALLY SIMPLE: DRAW THE LINE $\overline{AC}$, WHICH INTERSECTS BOTH **M** (AT C) AND $\overline{AB}$ (AT A). A LINE LIKE $\overline{AC}$ THAT INTERSECTS TWO OTHER LINES IS CALLED **TRANSVERSE** TO THEM, OR SIMPLY A **TRANSVERSAL.**

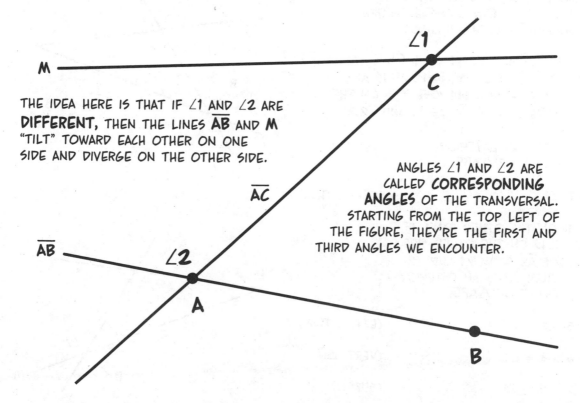

THE IDEA HERE IS THAT IF ∠1 AND ∠2 ARE **DIFFERENT,** THEN THE LINES $\overline{AB}$ AND **M** "TILT" TOWARD EACH OTHER ON ONE SIDE AND DIVERGE ON THE OTHER SIDE.

ANGLES ∠1 AND ∠2 ARE CALLED **CORRESPONDING ANGLES** OF THE TRANSVERSAL. STARTING FROM THE TOP LEFT OF THE FIGURE, THEY'RE THE FIRST AND THIRD ANGLES WE ENCOUNTER.

WE COULD ALSO START AT THE BOTTOM, OR ON THE OTHER SIDE OF $\overline{AC}$, SO THERE ARE SEVERAL PAIRS OF CORRESPONDING ANGLES MADE BY THIS TRANSVERSE LINE.

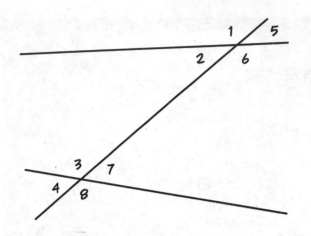

ONE PAIR OF CORRE-
SPONDING ANGLES ARE
EQUAL IF AND ONLY IF
ALL PAIRS ARE EQUAL.
FOR EXAMPLE, $\angle 2 = \angle 5$
AND $\angle 4 = \angle 7$ AS
VERTICAL ANGLES, SO
$\angle 2 = \angle 4$ IFF $\angle 5 = \angle 7$,
AND SO FORTH.

NOTE!

THE EXTERIOR ANGLE THEOREM HAS THIS IMPLICATION FOR INTERSECTING LINES:

Theorem 9-1. IF LINES L AND M INTERSECT, THEN ANY PAIR OF
CORRESPONDING ANGLES OF ANY TRANSVERSAL ARE **UNEQUAL.**

Proof. WE ASSUME THAT L
MEETS M AT A POINT P, AND WE
USE THE EXTERNAL ANGLE THEOREM
ON THE RESULTING TRIANGLE.

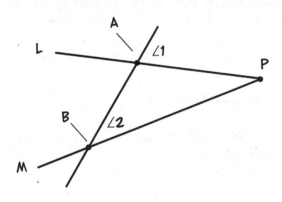

1. LET A BE A POINT ON L (A≠P), AND
   B A POINT ON M, B≠P. $\overline{AB}$ IS A
   TRANSVERSAL. SUPPOSE P IS ON THE
   SAME SIDE OF $\overline{AB}$ AS $\angle 1$ AND $\angle 2$.

2. $\angle 1$ IS AN EXTERIOR   (DEF.)
   ANGLE OF $\triangle ABP$.

3. $\angle 1 > \angle 2$ (EXT. $\angle$ THM.)

4. ON THE OTHER HAND, IF
   P IS ON THE OTHER SIDE
   OF $\overline{AB}$ FROM $\angle 1$ AND $\angle 2$,
   THEN $\angle 2$ IS AN EXTERIOR
   ANGLE OF $\triangle APB$.

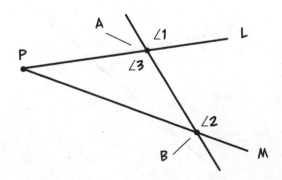

5. $\angle 2 > \angle 3$ (EXT. $\angle$ THM.)

6. $\angle 3 = \angle 1$ (VERT. $\angle$s)

7. $\angle 2 > \angle 1$ (SUBST.)

IN EITHER CASE, $\angle 1 \neq \angle 2$, AND
THE THEOREM IS PROVED. ▮

IT'S WORTH STATING THE **CONTRAPOSITIVE** OF THEOREM 9-1, WHICH MUST ALSO BE TRUE: IF A TRANSVERSAL MAKES **EQUAL** CORRESPONDING ANGLES, THEN LINES L AND M **DO NOT INTERSECT.** (IF THEY DID INTERSECT, THE ANGLES WOULD HAVE TO BE UNEQUAL.) MORE ON THIS SOON!

WHOA! NOT **EVER**???

AND WHAT ABOUT THE **CONVERSE** OF THEOREM 9-1? IF $\angle 1 \neq \angle 2$, SO THAT L AND M "TILT TOWARD" EACH OTHER, **MUST THEY INTERSECT?** AS PERSUASIVE AS THE DIAGRAM MAY BE, THIS CONVERSE **CANNOT BE PROVED** WITH OUR CURRENT SET OF ASSUMPTIONS.

OH, C'MON! THEY JUST **HAVE** TO!

DON'T THEY?

THAT DIDN'T STOP A LOT OF MATHEMATICIANS OVER THE CENTURIES FROM TRYING TO PROVE IT, BUT, ALAS, HISTORY IS A GRAVEYARD OF FRUSTRATED HOPES AND DISAPPOINTED DREAMS ABOUT THIS ONE.

Oh, well...

IT WAS ONLY IN THE 1800S, LONG AFTER EUCLID, THAT A TRIO OF MATHEMATICIANS UNDERSTOOD WHY.

GAUSS

LOBACHEVSKY

RIEMANN

EUCLID'S FIRST POSTULATES, AND OURS, ACTUALLY DESCRIBE **TWO DIFFERENT GEOMETRIES.**

ONE IS THE FAMILIAR KIND ON A FLAT PLANE.

THE OTHER, WHICH EUCLID NEVER DREAMED OF, OPERATES ON SADDLE-SHAPED SURFACES, WHERE "LINES" ARE CURVES THAT BEND AWAY FROM EACH OTHER AS THEY HEAD INTO THE DISTANCE.

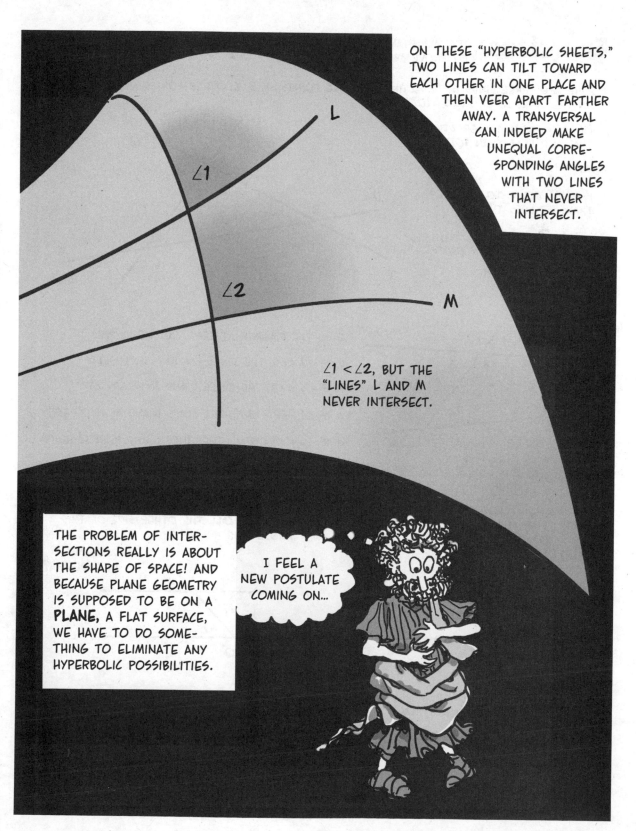

ON THESE "HYPERBOLIC SHEETS," TWO LINES CAN TILT TOWARD EACH OTHER IN ONE PLACE AND THEN VEER APART FARTHER AWAY. A TRANSVERSAL CAN INDEED MAKE UNEQUAL CORRE-SPONDING ANGLES WITH TWO LINES THAT NEVER INTERSECT.

$\angle 1 < \angle 2$, BUT THE "LINES" L AND M NEVER INTERSECT.

THE PROBLEM OF INTER-SECTIONS REALLY IS ABOUT THE SHAPE OF SPACE! AND BECAUSE PLANE GEOMETRY IS SUPPOSED TO BE ON A **PLANE,** A FLAT SURFACE, WE HAVE TO DO SOME-THING TO ELIMINATE ANY HYPERBOLIC POSSIBILITIES.

I FEEL A NEW POSTULATE COMING ON...

Exercises

IT'S IMPORTANT HERE TO DISTINGUISH BETWEEN WHAT WE **CAN** CONCLUDE OR PROVE ABOUT INTERSECTIONS AND WHAT WE STILL **CAN'T.** SO...

1. HERE ARE THREE LINES INTERSECTING AT THREE POINTS, A, B, C. OTHER POINTS ARE MARKED ON THE LINES ALSO. ∠PAQ=100°.

a. IS ∠ACR EQUAL TO, GREATER THAN, OR LESS THAN 100°?

b. WHAT IS ∠PAB?

c. IS ∠UBT EQUAL TO, GREATER THAN, OR LESS THAN 80°?

d. IS ∠QAC+∠ACR EQUAL TO, LESS THAN, OR MORE THAN 180°?

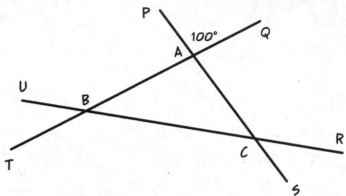

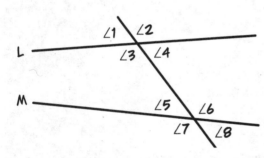

2. IN THE DIAGRAM AT LEFT, CAN WE PROVE:

a. IF ∠2<∠6, THE LINES L AND M INTERSECT?

b. IF ∠5 < ∠4, THE LINES L AND M INTERSECT?

c. IF ∠4+∠6>180°, THE LINES L AND M INTERSECT?

d. IF ∠2+∠8<180°, THE LINES L AND M INTERSECT?

3a. P IS A POINT NOT ON THE LINE L. DRAW A LINE FROM ANY POINT Q ON L THROUGH P. WHAT POSTULATE ALLOWS YOU TO DO THIS? SUPPOSE ∠PQR= 47°.

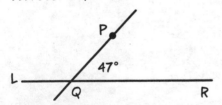

3b. COPY ∠PQR AT P ON THE SAME SIDE OF $\overline{PQ}$ AS ∠PQR. CALL THE OTHER SIDE OF THIS ANGLE LINE M.

3c. DOES M INTERSECT L?

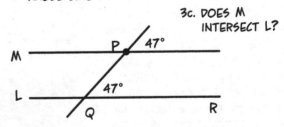

4. ASSUME THAT LINES V AND W **NEVER INTERSECT.** WHAT CAN WE PROVE?

a. ∠1=∠3

b. ∠6=∠7

c. ∠2+∠3=180°

d. ALL OF THE ABOVE

e. NONE OF THE ABOVE

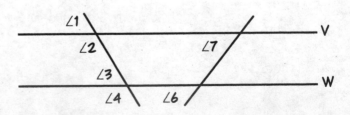

Chapter 10
BEING PARALLEL

Let's get right to the point—the point of intersection, that is. Let's **BANISH** that floppy surface by **POSTULATING** that our plane is **FLAT**. From now on, when lines approach each other, we **ASSUME** they intersect!

Postulate 10. IF A TRANSVERSAL MAKES **UNEQUAL** CORRESPONDING ANGLES WITH TWO LINES, THEN THE TWO LINES **INTERSECT**.

THIS POSTULATE ALLOWS US TO TEST WHETHER TWO LINES INTERSECT WITHOUT EVER LEAVING THE NEIGHBORHOOD: SIMPLY COMPARE TWO ANGLES ON THE PAGE IN FRONT OF US.

$\angle 1 \neq \angle 2 \Rightarrow$
L INTERSECTS M

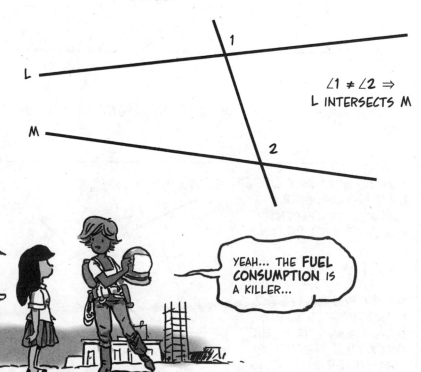

NO MORE WASTED TRIPS TO THE EDGE OF THE GALAXY!

YEAH... THE **FUEL CONSUMPTION** IS A KILLER...

LET'S SHIFT OUR POINT OF VIEW SLIGHTLY.
IF A POINT **P** IS OFF A LINE **L**, WHICH LINES
THROUGH P INTERSECT L?

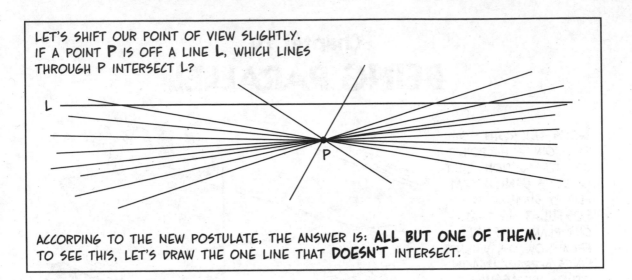

ACCORDING TO THE NEW POSTULATE, THE ANSWER IS: **ALL BUT ONE OF THEM.**
TO SEE THIS, LET'S DRAW THE ONE LINE THAT **DOESN'T** INTERSECT.

FIRST DRAW $\overline{PQ}$ THROUGH ANY POINT Q
ON L, MAKING AN ANGLE $\angle 1$.

COPY $\angle 1$ AT P MAKING $\angle 2 = \angle 1$ ON THE
SAME SIDE OF PQ AS $\angle 1$. CALL THE
NEW LINE **M**.

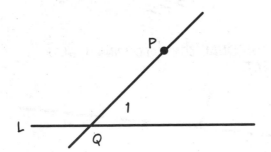

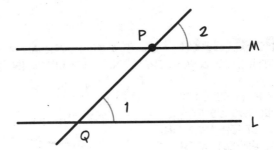

BY THEOREM 9-1, L AND M **DO NOT INTERSECT.** (IF THEY DID, THE ANGLES WOULD BE
UNEQUAL.)

NOW LOOK AT **ANY OTHER
LINE N** THROUGH P.
N MAKES A DIFFERENT
ANGLE $\angle 3$ WITH PQ, I.E.,

$$\angle 1 \neq \angle 3$$

SO, **BY POSTULATE 10,
N INTERSECTS L.** IN
OTHER WORDS, **M** IS THE
ONLY LINE THROUGH P
THAT **NEVER** INTERSECTS L.

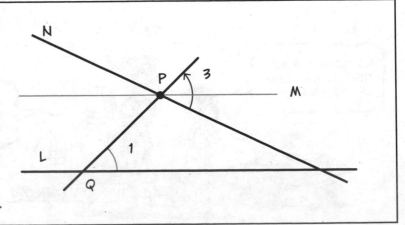

POSTULATE 10, THEN, IMPLIES THAT **ALL** LINES THROUGH P **EXCEPT FOR M** INTERSECT L.

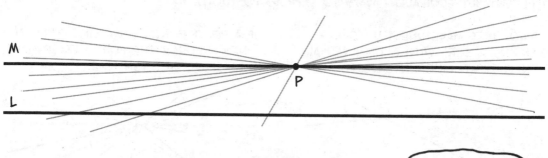

M

L

P

IN SUM, THERE IS **ONE AND ONLY ONE** LINE THROUGH P THAT NEVER TOUCHES L, A LINE THAT DESERVES ITS OWN NAME.

MAN, THAT IS ONE SPECIAL LINE...

SO, BASICALLY, INTERSECTING LINES ARE COMMON AS DIRT NOW.

HEY! DIRT IS **DELICIOUS!**

Definition. TWO LINES IN THE SAME PLANE ARE **PARALLEL** IFF THEY DO NOT INTERSECT. WE SAY "L IS PARALLEL TO M" OR "L AND M ARE PARALLEL" AND WRITE:

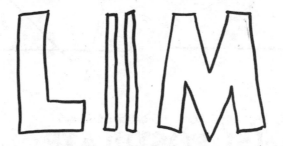

L ‖ M

IN THESE TERMS, POSTULATE 10 SAYS: IF TWO LINES ARE PARALLEL, THEN **EVERY TRANSVERSAL MAKES EQUAL CORRESPONDING ANGLES.**

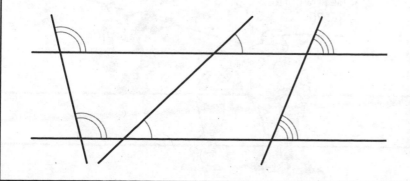

IF EVEN **ONE PAIR** OF CORRESPONDING ANGLES IS UNEQUAL, THE LINES MUST INTERSECT! (AND EVERY OTHER TRANSVERSAL MUST ALSO MAKE UNEQUAL CORRESPONDING ANGLES.)

THE REASONING ON THE PREVIOUS TWO PAGES SHOWS THAT THERE ARE SEVERAL DIFFERENT BUT EQUIVALENT WAYS TO EXPRESS POSTULATE 10.

IF TWO LINES MAKE UNEQUAL CORRE-SPONDING ANGLES WITH A TRANSVERSAL, THEN THE TWO LINES INTERSECT.

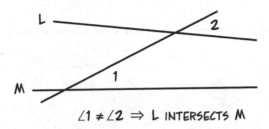

$$\angle 1 \neq \angle 2 \Rightarrow L \text{ INTERSECTS } M$$

IF A POINT IS NOT ON A LINE, THEN ALL LINES THROUGH THE POINT, EXCEPT ONE, INTERSECT THE GIVEN LINE.

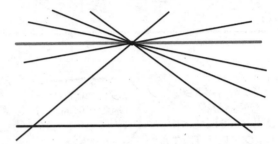

IF A POINT IS NOT ON A LINE, THEN ONE AND ONLY ONE LINE THROUGH THE POINT IS PARALLEL TO THE LINE.

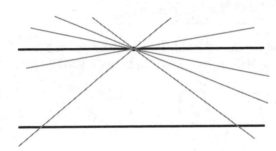

IF TWO LINES ARE PARALLEL, THEN EVERY TRANSVERSAL INTERSECTS THEM WITH EQUAL CORRESPONDING ANGLES.

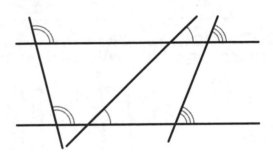

WE REFER TO ANY ONE OF THESE FOUR FORMULATIONS AS

THE PARALLEL POSTULATE.

(IT COULD JUST AS WELL BE CALLED "THE INTERSECTION POSTULATE" BUT NEVER IS.)

PARALLEL LINES GET ALL THE RESPECT...

Examples

SUPPOSE L∥M. WHAT IS ∠1?

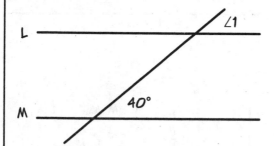

∠1 = 40°. BY THE PARALLEL POSTU-
LATE, PARALLEL LINES MAKE EQUAL
CORRESPONDING ANGLES.

SUPPOSE THE ANGLES ARE AS MARKED.
DO L AND M INTERSECT?

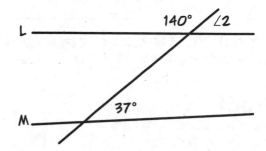

YES. THE 140° ANGLE FORMS A LINEAR
PAIR WITH ∠2, SO ∠2=180°-140°= 40°.
THE CORRESPONDING ANGLE IS 37°, AND
37° ≠ 40°.

HERE'S A SIMPLE AND WELCOME CONSEQUENCE:

Theorem 10-1. GIVEN THREE LINES IN A PLANE, IF TWO LINES ARE PARALLEL
TO THE THIRD, THEN THEY ARE PARALLEL TO EACH OTHER.

Proof. ASSUMING THAT L∥M AND M∥N,
WE SHOW THAT L∥N.

1. LET **A** BE ANY POINT   (RULER POST.)
   ON L AND **B** ANY
   POINT ON M.

2. THE LINE $\overline{AB}$ IS DIF-   (POST. 2)
   FERENT FROM M BE-
   CAUSE **A** IS NOT ON M.

3. M IS THE ONLY LINE   (PAR. POST.)
   THROUGH **B** PARALLEL
   TO N, SO $\overline{AB}$ INTER-
   SECTS N AT A POINT **C**.

4. ∠1 = ∠2 AND ∠2 = ∠3   (PAR. POST.)

5. ∠1 = ∠3   (SUBST.)

6. L∥N ■   (THM. 9-1)

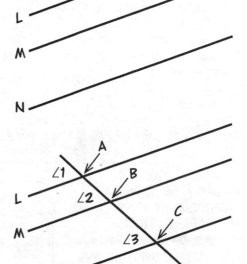

Parallels and Perpendiculars

THE PARALLEL POSTULATE ENSURES THAT PERPEN-
DICULARS AND PARALLELS PLAY NICE WITH EACH OTHER.

Theorem 10-2. ALL PERPENDICULARS TO A
GIVEN LINE ARE PARALLEL TO EACH OTHER.

Proof. PERPENDICULARS
ALL MAKE CORRESPONDING
ANGLES OF 90°. EQUAL
ANGLES IMPLY PARALLELISM
BY THEOREM 9-1. ▪

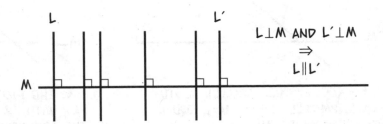

$L \perp M$ AND $L' \perp M$
$\Rightarrow$
$L \| L'$

Theorem 10-3. IF TWO LINES ARE PARALLEL, THEN A PERPENDICULAR TO ONE
IS PERPENDICULAR TO THE OTHER.

Proof. ASSUME L, L', AND M ARE LINES IN A
PLANE, WITH $L \| L'$ AND $L \perp M$. WE SHOW $L' \perp M$.

1. M CAN'T BE PARALLEL TO L'.   (PAR. POST.)
IF IT WERE, THEN L AND M
WOULD BE TWO LINES
THROUGH A PARALLEL TO L'.

2. L' INTERSECTS M AT SOME   (DEF. OF
POINT **B**, MAKING ∠2 COR-   PARALLEL)
RESPOND TO ∠1.

3. ∠1 = ∠2   (PAR. POST.)

4. ∠2 = 90°, AND $L' \perp M$. ▪   (DEF. OF PERP.)

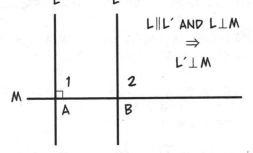

$L \| L'$ AND $L \perp M$
$\Rightarrow$
$L' \perp M$

NOTE THE USE
OF THE PARALLEL
POSTULATE IN
STEPS 1 AND 3!

WE CAN ALSO EXPRESS
THESE THEOREMS IN TERMS
OF **INTERSECTIONS**
INSTEAD OF PARALLELISM.
THAT MEANS TWISTING THE
STATEMENTS INTO THEIR
CONTRAPOSITIVES (AS
EXPLAINED ON P. 25).

IF I SAY THE
MAGIC WORD,
THIS RABBIT
WILL EXPLAIN
EVERYTHING!

HM... A MUTE
RABBIT... YOU MUST
NOT HAVE SAID
THE MAGIC WORD...

THEOREM 10-2 SAYS: IF L INTERSECTS L' (THEY'RE **NOT** PARALLEL) AND L⊥M, THEN L' IS **NOT** PERPENDICULAR TO M.

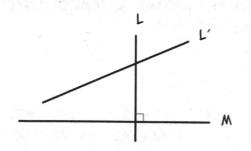

THEOREM 10-3 SAYS: IF L' IS **NOT** PERPENDICULAR TO M AND L⊥M, THEN L INTERSECTS L' (THEY'RE **NOT** PARALLEL).

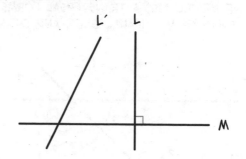

THESE IMPLY THE FOUNDATION FOR THEOREM 8-1. LET'S CALL IT:

Theorem 8-0. PERPENDICULARS TO TWO SIDES OF A TRIANGLE INTERSECT.

Proof. SUPPOSE L⊥AB, AND L'⊥BC. WE SHOW THAT L INTERSECTS L'.

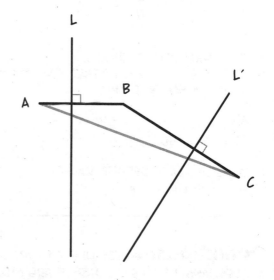

1. AB ∦ BC   (AB, BC INTERSECT AT B, FORMING AN ANGLE <180°.)

2. AB⊥L   (ASSUMED)

3. L IS NOT PERPEN-DICULAR TO BC.   (THM. 10-2)

4. L'⊥BC   (ASSUMED)

5. L INTERSECTS L'. ■ (THM. 10-3)

THIS FINALLY PLUGS THE HOLE IN THE CONSTRUCTION ON PAGE 87. THERE, WE COULDN'T BE SURE THAT THESE TWO PERPENDICULARS INTERSECTED. NOW WE SEE THAT IT TAKES THE PARALLEL POSTULATE TO PROVE THAT THEY DO!

PARALLEL POSTULATE!

AN IMPLICATION P⇒Q HAS AN EQUALLY VALID CONTRA-POSITIVE ~Q ⇒ ~P; POSTU-LATE 10 (OR ITS CONTRA-POSITIVE) VALIDATES ALL GEOMETRIC CONSTRUCTIONS BY GUARANTEEING THAT LINES INTERSECT WHEN THEY "SHOULD." GET IT?

THE MAGIC WORDS!

Other Transversal Angles

AS WE MENTIONED, A TRANSVERSAL FORMS A TOTAL OF EIGHT ANGLES. SOME OF THESE ARE PAIRS OF VERTICAL ANGLES, AND OTHERS ARE LINEAR PAIRS.

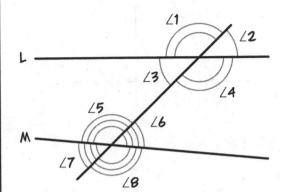

$\angle 1 = \angle 4$ $\angle 1 + \angle 2 = \angle 1 + \angle 3 = 180°$

$\angle 2 = \angle 3$ $\angle 2 + \angle 4 = \angle 3 + \angle 4 = 180°$

$\angle 5 = \angle 8$ $\angle 5 + \angle 6 = \angle 5 + \angle 7 = 180°$

$\angle 6 = \angle 7$ $\angle 6 + \angle 8 = \angle 7 + \angle 8 = 180°$

BY THE PARALLEL POSTULATE, IF L∥M, THEN $\angle 3 = \angle 7$. IN THAT CASE, IT'S ALSO TRUE THAT

$\angle 3 = \angle 6$

$\angle 3 + \angle 5 = 180°$

BECAUSE OF ALL THOSE VERTICAL ANGLES AND LINEAR PAIRS.

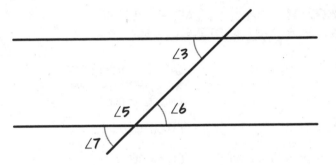

Definitions. ANGLES LIKE $\angle 3$ AND $\angle 6$ ARE CALLED **ALTERNATING INTERIOR** ANGLES. ANGLES LIKE $\angle 3$ AND $\angle 5$ ARE CALLED **ADJACENT INTERIOR** ANGLES.

AND WE SEE THAT TWO LINES ARE PARALLEL IFF

CORRESPONDING ANGLES ARE EQUAL

ALTERNATING INTERIOR ANGLES ARE EQUAL

ADJACENT INTERIOR ANGLES ARE SUPPLEMENTARY.

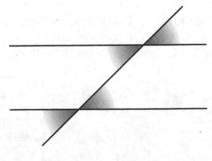

ALL ∠ ARE EQUAL; ALL ◢ ARE EQUAL; ANY ∠ PLUS ANY ◢ EQUALS 180°.

HOW DO YOU PRONOUNCE "◢"?

AND NOW, COMPLETELY OUT OF THE BLUE...

IT'S NOT A BIRD!

IT'S NOT A PLANE!

IT **IS** KIND OF A SUPER-THEOREM, THOUGH...

Theorem 10-4. THE ANGLES IN ANY TRIANGLE ADD TO **180°**.

Proof.

1. GIVEN △ABC, DRAW LINE L THROUGH A WITH ∠1 = ∠C.
   (COPY ANGLE)

2. L∥BC
   (THM. 9-1)

3. ∠2 = ∠B
   (PAR. POST.)

4. ∠1 + ∠3 + ∠2 = 180°
   (LINEAR TRIPLE)

5. ∠B + ∠3 + ∠C = 180° ▮
   (SUBST.)

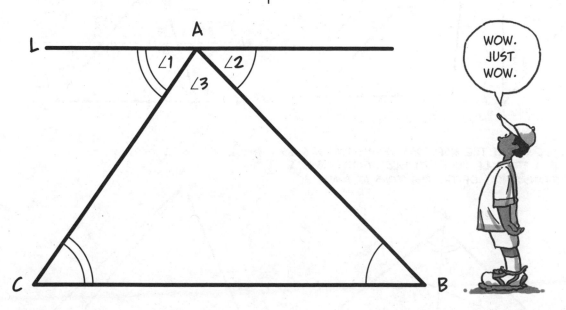

WOW. JUST WOW.

SURPRISING, ISN'T IT? A POSTULATE ABOUT INTERSECTING LINES ENDS UP GIVING AN **EXACT MEASURE** FOR EVERY TRIANGLE.

Exercises

1a. DO L AND M INTERSECT? IF SO, AT WHAT ANGLE?

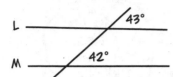

1b. L∥M. WHAT IS ∠2?

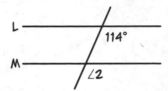

2. IF AB∥CD AND AD∥BC, SHOW THAT △ABD ≅ △CDB.

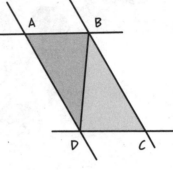

1c. L AND M ARE NOT PARALLEL. WHICH IS GREATER, ∠1 OR ∠2?

1d. L∥M. WHAT IS ∠3?

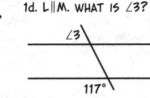

3. FIND THE MISSING ANGLE IN EACH TRIANGLE.

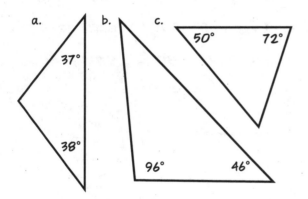

4. L, M, AND N ARE THREE PARALLEL LINES.

a. SHOW THAT ∠9 + ∠10 = 180° + ∠2

b. IF ∠10=128° AND ∠13=73°, WHAT ARE ALL THE OTHER ANGLES?

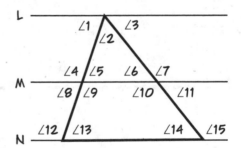

5. SHOW THAT THE BISECTORS OF ANY TWO ANGLES IN A TRIANGLE MUST INTERSECT. (HINT: AB IS A TRANSVERSAL OF THE BISECTORS AT A AND B.)

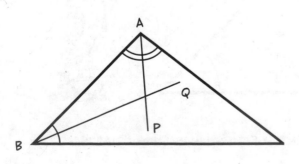

6. WHY IS ∠B < ∠ADC?

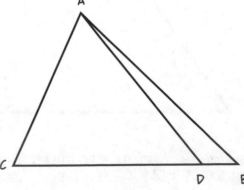

Chapter 11
TRIANGLES IN A FLAT WORLD

WITH THE PARALLEL POSTULATE AND 180° TRIANGLES, LIFE GETS MUCH EASIER.
IF WE KNOW ANY TWO ANGLES OF A TRIANGLE, WE KNOW THE THIRD!

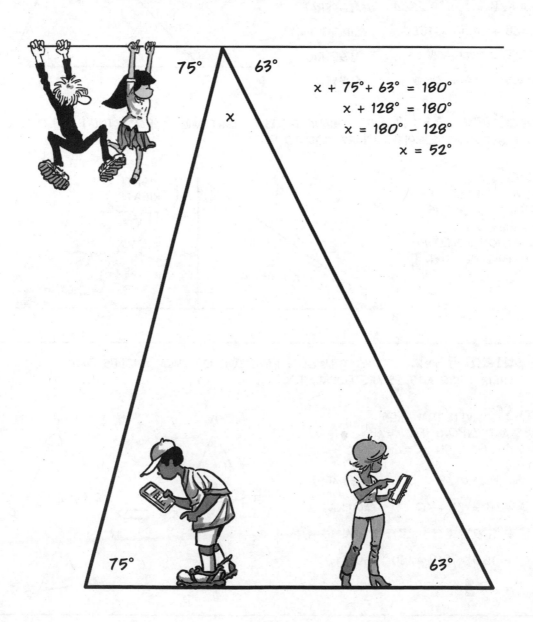

75°   63°

x

$$x + 75° + 63° = 180°$$
$$x + 128° = 180°$$
$$x = 180° - 128°$$
$$x = 52°$$

75°   63°

CONSEQUENTLY:

Theorem 11-1. AN EXTERNAL ANGLE OF A TRIANGLE EQUALS THE SUM OF THE TWO REMOTE ANGLES.

Proof. GIVEN $\triangle ABC$ WITH REMOTE ANGLE $\angle ACD$, WE PROVE THAT $\angle ACD = \angle A + \angle B$.

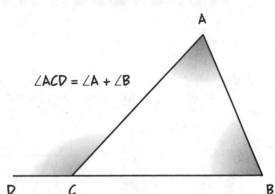

$\angle ACD = \angle A + \angle B$

1. $\angle A + \angle B + \angle ACB = 180°$ (THM. 10-4)

2. $\angle A + \angle B = 180° - \angle ACB$ (ALGEBRA)

3. $\angle ACB + \angle ACD = 180°$ (LINEAR PAIR)

4. $\angle ACD = 180° - \angle ACB$ (ALGEBRA)

5. $\angle ACD = \angle A + \angle B$ ∎   (SUBST.)

Corollary 11-1.1. IF A TRIANGLE HAS A RIGHT ANGLE (90°), THEN THE OTHER TWO ANGLES ARE COMPLEMENTARY (ADD TO 90°).

Proof. AN EXTERNAL ANGLE AT THE RIGHT ANGLE IS ALSO RIGHT. THE CONCLUSION FOLLOWS FROM THEOREM 11-1. ∎

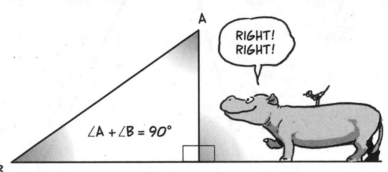

$\angle A + \angle B = 90°$

RIGHT! RIGHT!

Theorem 11-2. IF TWO TRIANGLES HAVE TWO OF THEIR ANGLES EQUAL, THEN THEIR THIRD ANGLES ARE EQUAL, TOO.

Proof. GIVEN TRANGLES $\triangle ABC$ AND $\triangle PQR$, IF $\angle A = \angle P$ AND $\angle B = \angle Q$, THEN $\angle C = \angle R$.

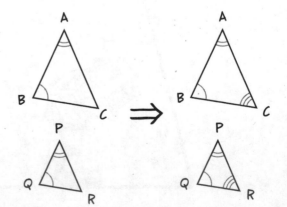

1. $\angle A = \angle P$ AND $\angle B = \angle Q$ (ASSUMED)

2. $\angle A + \angle B = \angle P + \angle Q$ (ADDING)

3. $\angle C = 180° - (\angle A + \angle B)$ (THM. 10-4)

4. $\quad = 180° - (\angle P + \angle Q)$ (SUBST.)

5. $\quad = \angle R$ ∎   (THM. 10-4)

HERE'S ANOTHER ONE THAT FLOWS FROM THEOREM 10-4, AND ULTIMATELY, FROM THE PARALLEL POSTULATE.

Theorem 11-3 (AAS). IF A TRIANGLE HAS TWO ANGLES EQUAL TO TWO ANGLES OF ANOTHER TRIANGLE, AND **ANY SIDE** OF ONE EQUALS THE CORRESPONDING SIDE OF ANOTHER, THEN THE TRIANGLES ARE CONGRUENT.

Proof. SINCE TWO ANGLES ARE EQUAL, THE THIRD ANGLES MUST ALSO BE EQUAL, BY THEOREM 11-2. ASSUMING THAT $\triangle ABC$ AND $\triangle PQR$ HAVE $\angle A = \angle P$, $\angle B = \angle Q$, AND $BC = QR$, WE SHOW THAT $\triangle ABC \cong \triangle PQR$.

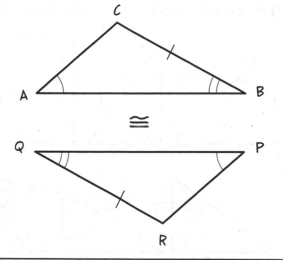

1. $\angle A = \angle P$, $\angle B = \angle Q$ (ASSUMED)

2. SO $\angle C = \angle R$ (THM. 11-2)

3. $BC = QR$ (ASSUMED)

4. $\triangle ABC \cong \triangle PQR$ ▮   (ASA)

AK! WHAT IS THAT... THAT... **ANIMAL** DOING HERE?

SOMETHING TO DO WITH RIGHT TRIANGLES, I IMAGINE...

Triangle Gallery

WE CAN NOW CONFIDENTLY
IDENTIFY TRIANGLE **SHAPES**
BY THEIR THREE ANGLES.

ANY **EQUILATERAL** TRIANGLE—
ONE WITH THREE EQUAL **SIDES**—
ALSO HAS THREE EQUAL **ANGLES**,
EACH EQUAL TO $\frac{1}{3}(180°)$ OR **60°**
(REGARDLESS OF THE LENGTH OF
THE SIDES).

IT'S ISOSCELES THREE WAYS!

60°

60°   60°

RIGHT TRIANGLES—THOSE WITH A RIGHT ANGLE—COME IN AN ASSORTMENT OF SHAPES.

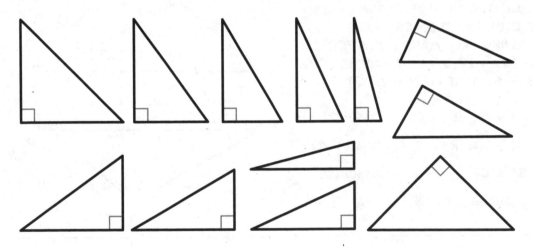

HERE'S A SPECIAL
ONE: BISECT THE
VERTEX ANGLE OF
AN EQUILATERAL
TRIANGLE. THIS LINE
IS ALSO THE PERPEN-
DICULAR BISECTOR OF
THE OPPOSITE SIDE,
BY THEOREM 6-3.

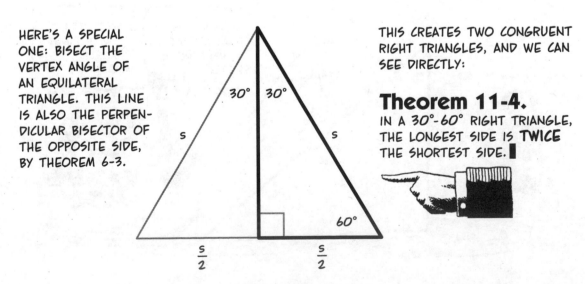

30°   30°

s   s

$\frac{s}{2}$ $\frac{s}{2}$

60°

THIS CREATES TWO CONGRUENT
RIGHT TRIANGLES, AND WE CAN
SEE DIRECTLY:

Theorem 11-4.
IN A 30°-60° RIGHT TRIANGLE,
THE LONGEST SIDE IS **TWICE**
THE SHORTEST SIDE. ∎

ON PAGE 79, WE DEFINED THE **HYPOTENUSE** OF A RIGHT TRIANGLE AS ITS LONGEST SIDE, THE SIDE OPPOSITE THE RIGHT ANGLE.

NOT THE HIPPOPOTANUSE OR THE HYPOPOTAMUS OR THE POPOHIPPENUPS; IT'S THE *HIGH-PAH-TEN-USE.*

RIGHT TRIANGLES HAVE THEIR VERY OWN CONGRUENCE TEST:

Theorem 11-5. IF THE HYPOTENUSE AND ONE LEG OF A RIGHT TRIANGLE ARE EQUAL TO THE HYPOTENUSE AND LEG OF ANOTHER RIGHT TRIANGLE, THEN THE TRIANGLES ARE CONGRUENT.

Proof. LET THE TRIANGLES BE △ABC AND △PQR, WITH AC = PR, AB = PQ, AND ∠C = ∠R = 90°.

1. CONSTRUCT △ACD ≅ △PRQ ON AC, ON THE OPPOSITE SIDE FROM △ABC. THIS WORKS BECAUSE AC = PR.

2. ∠BCA = ∠DCA = 90°, SO ∠BCA + ∠DCA = 180°; THEN BCD IS A SEGMENT, AND ABCD IS A TRIANGLE.

3. SINCE AB = AD, △ABD IS ISOSCELES, BY DEFINITION.

4. ∠B = ∠D, BY THM. 6-1.

5. △ACB ≅ △ACD, BY AAS.

6. △ACB ≅ △PRQ, AS BOTH ARE CONGRUENT TO △ACD. ∎

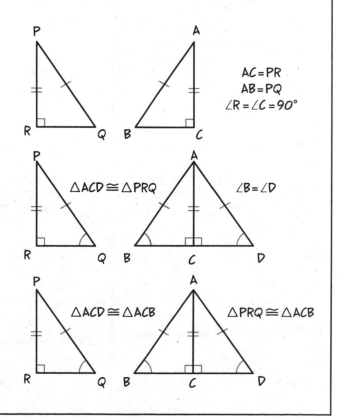

Example: A Flat Mirror

WHEN KEVIN STANDS IN FRONT OF A MIRROR, HE SEES HIS REFLECTION AS IF IT WERE AT A POINT BEHIND THE MIRROR. WHERE IS THAT POINT?

LET'S VIEW THE SCENE EDGE-ON FROM ABOVE, A PERSPECTIVE FROM WHICH THE MIRROR LOOKS LIKE A VERTICAL LINE.

THE MIRROR BOUNCES LIGHT OFF ITS SURFACE. SOME BOUNCING LIGHT RAYS BOUNCE INTO KEVIN'S EYE AT **K**. LUCKILY, THIS DOESN'T HURT.

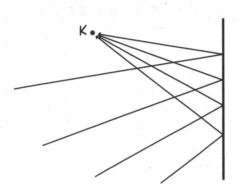

IT'S KNOWN THAT ANY LIGHT RAY MAKES THE SAME ANGLE COMING OFF THE MIRROR AS IT MADE GOING IN, TRADITIONALLY MEASURED AGAINST A PERPENDICULAR **L** TO THE MIRROR.

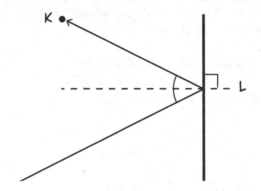

THIS LIGHT RAY LEAVES AN OBJECT AT P, HITS THE MIRROR AT Q, AND COMES TO KEVIN'S EYE AT K.

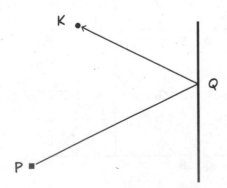

DRAW A LINE PERPENDICULAR TO THE MIRROR FROM P, INTERSECTING THE MIRROR AT R. PR IS THE DISTANCE FROM P TO THE MIRROR (AS SHOWN ON P. 79).

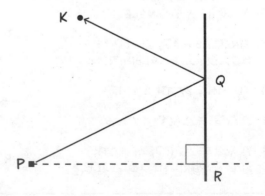

114

EXTENDING KEVIN'S LINE OF SIGHT KQ CREATES TWO RIGHT TRIANGLES △QPR AND △QSR.

THE OBJECT "SEES ITSELF" SOMEWHERE ALONG RS, WHILE KEVIN SEES IT ALONG QS, SO THE REFLECTION IS AT THE LINES' INTERSECTION S.

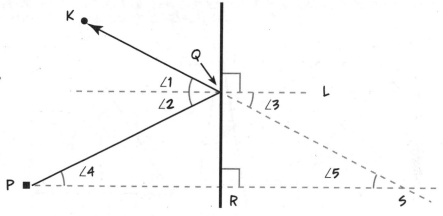

WE NOW PROVE THAT PR = RS.

1. ∠1 = ∠2	(ASSUMED)		6. ∠2 = ∠4, ∠3 = ∠5	(PAR. POST.)
2. ∠1 = ∠3	(VERT. ∠s)		7. ∠4 = ∠5	(SUBST.)
3. ∠2 = ∠3	(SUBST.)		8. QR = QR	
4. L⊥QR, PS⊥QR	(ASSUMED)		9. △QPR ≅ △QSR	(AAS)
5. L∥PS	(THM. 10-2)		10. **PR = RS** ▮	(CORR. PARTS)

A MIRROR IMAGE IS **THE SAME DISTANCE BEHIND THE MIRROR** AS THE OBJECT IS IN FRONT OF IT. THIS EXPLAINS WHY A MIRRORED WALL SEEMS TO DOUBLE THE SIZE OF A ROOM.

Exercises

1. SUPPLY THE MISSING ANGLE.

a.

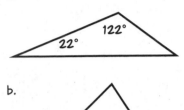

b.

2. SUPPLY THE MISSING ANGLES IN THESE ISOSCELES TRIANGLES.

a.

b.

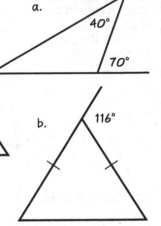

3. SUPPLY THE MISSING INTERNAL ANGLES.

a.

b.

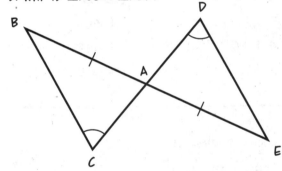

4. WHAT IS THE SUM OF THE THREE EXTERNAL ANGLES OF A TRIANGLE?

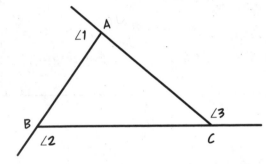

5. WHY IS △ABC≅△AED?

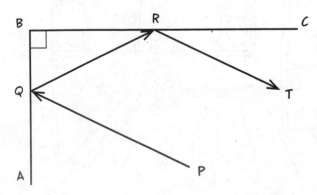

6. TWO MIRRORS AB AND BC ARE JOINED AT RIGHT ANGLES. A RAY OF LIGHT COMES FROM P, STRIKES AB AT Q, REFLECTS TOWARD BC, HITS IT AT R, AND REFLECTS TOWARD T. SHOW THAT PQ∥RT.

7. WAIT A MINUTE! DOESN'T THIS MEAN THAT AN OB-SERVER AT P **CAN'T SEE HIS OWN REFLECTION** FROM MIRROR AB IN MIR-ROR BC? THE RAY RT NEVER GETS CLOSE TO AN EYE AT P.

AND YET EXPERIENCE TELLS US THAT WE **CAN** SEE OUR REFLECTIONS FROM ONE MIRROR IN ANOTHER. HOW DO YOU EXPLAIN THIS?

Chapter 12
ONE MORE SIDE!

'TIS TIME, SAID THE GEOMETER,
TO SPEAK OF SOMETHING MORE:
MORE ANGLES, SIDES, MORE EVERYTHING,
UP TO THE NUMBER FOUR.
—NOT LEWIS CARROLL

ALTHOUGH WE HAVE DEVOTED MANY PAGES TO TRIANGLES, THE WORLD AROUND US, AT LEAST VIEWED FROM MY CHAIR, IS FULL OF FOUR-SIDED OBJECTS, INCLUDING THIS PIECE OF PAPER.

IN THIS CHAPTER, WE ADD A NEW SIDE TO OUR GEOMETRICAL—AND PRACTICAL—KNOWLEDGE, WITH ADVICE ABOUT FRAMES, WALLS, AND OTHER CONSTRUCTION PROJECTS.

LET'S THINK INSIDE THE BOX!

FIRST, WE HAVE TO
DEFINE WHAT WE'RE
TALKING ABOUT.

YES!

Definition. A QUADRILATERAL CONSISTS OF FOUR POINTS, CALLED **VERTICES**, AND FOUR LINE SEGMENTS, OR SIDES. EACH SEGMENT HAS TWO OF THE VERTICES AS ENDPOINTS; EVERY VERTEX BELONGS TO EXACTLY TWO SIDES; AND THE SIDES INTERSECT NOWHERE ELSE.

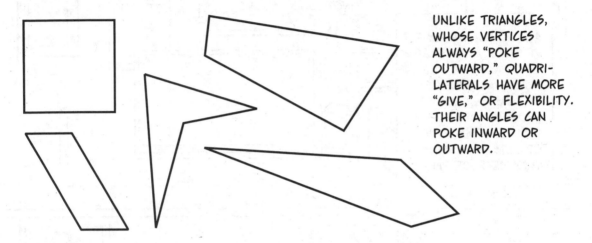

UNLIKE TRIANGLES, WHOSE VERTICES ALWAYS "POKE OUTWARD," QUADRILATERALS HAVE MORE "GIVE," OR FLEXIBILITY. THEIR ANGLES CAN POKE INWARD OR OUTWARD.

IN THIS BOOK, WE'LL DEAL ALMOST EXCLUSIVELY WITH POKING-OUT, OR **CONVEX**, FIGURES. EVERY INTERIOR ANGLE WILL BE LESS THAN 180°.

ANOTHER WAY TO THINK OF CONVEXITY IS IN TERMS OF THE FIGURE'S INSIDE AND OUTSIDE.* IF TWO POINTS ARE INSIDE A CONVEX FIGURE, THEN SO IS THE WHOLE LINE SEGMENT BETWEEN THEM.

THEN HOW DID WE GET OUT?

3-D, BABY, THREE-DEE!

*THE FACT THAT A QUADRILATERAL HAS AN INSIDE AND AN OUTSIDE IS PROVED IN MORE ADVANCED COURSES.

A SEGMENT JOINING NONADJACENT OR OPPOSITE VERTICES, LIKE **BD** HERE, LIES INSIDE OUR CONVEX QUADRILATERIAL. SUCH A SEGMENT IS CALLED A **DIAGONAL.**

A DIAGONAL DIVIDES THE QUADRILATERAL INTO TWO TRIANGLES. CLEARLY, THE QUADRILATERAL'S ANGLES TOTAL THE ANGLES OF THESE TWO TRIANGLES.

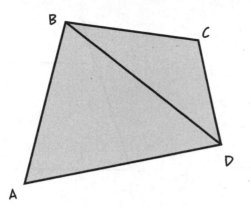

WHICH PROVES THIS:

Theorem 12-1.
A CONVEX QUADRILATERAL'S ANGLES ADD UP TO 360°. ∎

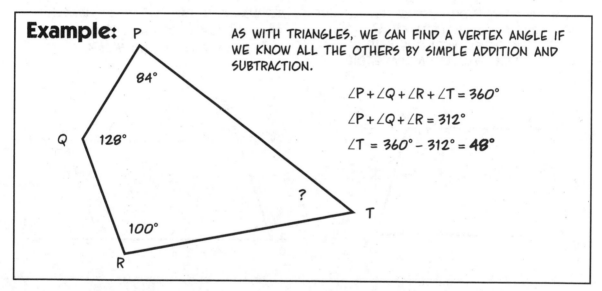

Example:

AS WITH TRIANGLES, WE CAN FIND A VERTEX ANGLE IF WE KNOW ALL THE OTHERS BY SIMPLE ADDITION AND SUBTRACTION.

$$\angle P + \angle Q + \angle R + \angle T = 360°$$

$$\angle P + \angle Q + \angle R = 312°$$

$$\angle T = 360° - 312° = \mathbf{48°}$$

P

84°

Q

128°

?

T

100°

R

IT'S THIS FULL 360° THAT GIVES QUADRI-LATERALS SO MUCH VARIETY. UNLIKE TRIANGLES, QUADRI-LATERALS CAN HAVE **PARALLEL SIDES**— WHICH EXPLAINS THEIR BIG ROLE IN THE BUILDING TRADES!

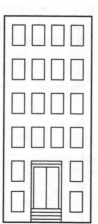

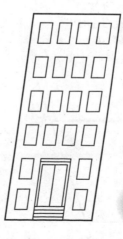

FINE JOB, PEOPLE! NOW WE LET THE INSURANCE TRADES HAVE A TURN!

IT'S EASY TO CONSTRUCT QUADRILATERALS WITH PARALLEL SIDES. START WITH ANY TWO SEGMENTS AB AND BC THAT MEET AT AN ANGLE LESS THAN 180°.

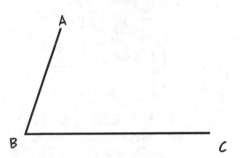

ON THE RAY $\overrightarrow{BA}$, COPY ∠B AT POINT A TO MAKE A RAY $\overrightarrow{L}$ WITH $\overrightarrow{L} \parallel BC$.

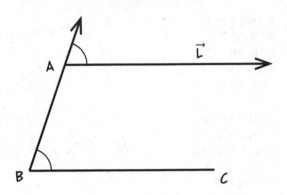

FROM ANY POINT D ON $\overrightarrow{L}$, DRAW THE SEGMENT CD, MAKING A QUADRILATERAL **ABCD** WITH AD∥BC. THIS FIGURE, WITH **ONE PAIR** OF PARALLEL SIDES, IS CALLED A **TRAPEZOID**.

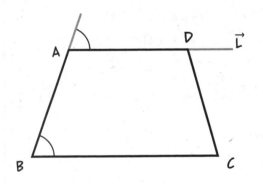

TO MAKE BOTH PAIRS OF SIDES PARALLEL, DRAW THE RAY $\overrightarrow{BC}$ BEYOND C AND COPY ∠B AT C TO MAKE RAY $\overrightarrow{M}$ WITH $\overrightarrow{M} \parallel AB$.

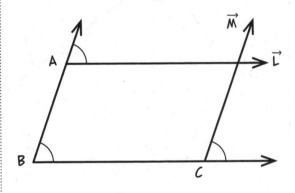

$\overrightarrow{L}$ AND $\overrightarrow{M}$ INTERSECT AT A POINT E, MAKING A QUADRILATERAL **ABCE** WITH TWO PAIRS OF PARALLEL SIDES: A **PARALLELOGRAM.**

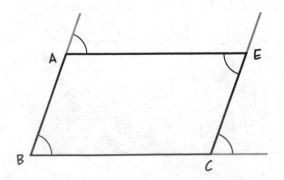

IF ∠B HAPPENED TO BE 90°, THIS PARALLELOGRAM WOULD HAVE FOUR EQUAL RIGHT ANGLES, AND IT WOULD BE A **RECTANGLE.**

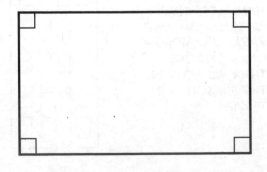

THIS FOLLOWS FROM THE CONSTRUCTION:

Theorem 12-2. IN A PARALLELOGRAM, OPPOSITE ANGLES ARE EQUAL.

Proof. EACH SIDE IS A TRANSVERSAL OF TWO PARALLEL LINES. TWO ADJACENT ANGLES, BEING ADJACENT INTERIOR ANGLES OF THE TRANSVERSAL, SUM TO 180°, BY THE PARALLEL POSTULATE.

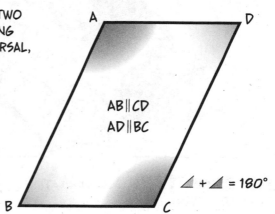

1. $\angle A + \angle B = 180°$ (BECAUSE AD∥BC)

2. $\angle B + \angle C = 180°$ (BECAUSE AB∥CD)

3. $\angle A - \angle C = 0°$ (SUBTRACTION)

4. $\angle A = \angle C$ (ALGEBRA)

5. SIMILARLY, $\angle B = \angle D$ ▮

CONVERSELY:

Theorem 12-3. IF A QUADRILATERAL HAS BOTH PAIRS OF OPPOSITE ANGLES EQUAL, THEN IT'S A PARALLELOGRAM.

Proof. GIVEN ABCD WITH $\angle A = \angle C$, $\angle B = \angle D$

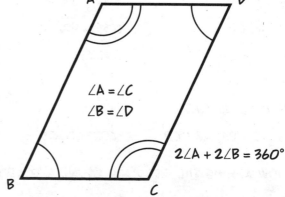

1. $\angle A + \angle B + \angle C + \angle D = 360°$ (THM. 12-1)

2. $\angle A = \angle C$, $\angle B = \angle D$ (ASSUMED)

3. $\angle A + \angle B + \angle A + \angle B = 360°$ (SUBST.)

4. $2(\angle A + \angle B) = 360°$

5. $\angle A + \angle B = 180°$

6. AD∥BC   (THM. 9-1)

7. SIMILARLY, $\angle A + \angle D = 180°$

8. AB∥CD ▮   (THM. 9-1)

IF ONLY **ONE** PAIR OF OPPOSITE ANGLES IS EQUAL, IT'S NO PARALLELOGRAM.

NOW, AT LAST, A THEOREM FOR CARPENTERS:

I JUST **KNEW** THIS BOOK WAS GOOD FOR SOMETHING BESIDES DRIVING NAILS!

Theorem 12-4. A QUADRILATERAL IS A PARALLELOGRAM IFF ITS OPPOSITE SIDES ARE EQUAL.

Proof. FIRST ASSUME AB∥CD, AD∥BC. WE SHOW EQUALITY OF THE PARALLEL SIDES BY DIVIDING THE PARALLELOGRAM INTO CONGRUENT TRIANGLES.

1. DRAW DIAGONAL AC.   (POST. 2)

2. AD∥BC   (ASSUMED)

3. ∠1 = ∠2   (PAR. POST.)

4. AB∥CD   (ASSUMED)

5. ∠3 = ∠4   (PAR. POST.)

6. AC = AC

7. △ABC ≅ △CDA   (ASA)

8. AD = BC, AB = CD   (CORR. PARTS)

NOW ASSUME THE SIDES ARE EQUAL, AND WE PROVE THEM PARALLEL.

9. △ABC ≅ △CDA   (SSS)

10. ∠1 = ∠2   (CORR. PARTS)

11. AB∥CD   (THM. 9-1)

12. ∠3 = ∠4   (CORR. PARTS)

13. AD∥BC   (THM. 9-1)

14. ABCD IS A PARALLELOGRAM.   (DEFINITION) ∎

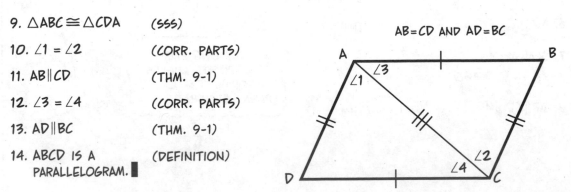

TO MAKE A FRAME WITH PARALLEL SIDES, ALL YOU HAVE TO DO IS CUT EQUAL SIDES, AND PARALLELISM IS GUARANTEED!

SO KEVIN MAKES A PARALLELOGRAM.

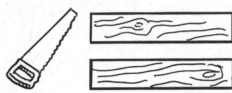

YES!

TOO BAD PARALLELOGRAM SHAPES TEND TO FLOP OVER.

CRASH

HOW DOES KEVIN MAKE A BOOKSHELF **RECTANGULAR** AND KEEP IT THAT WAY?

OH, WELL, AT LEAST EVERYTHING IS STILL PARALLEL...

Theorem 12-5. A PARALLELOGRAM IS A RECTANGLE IFF ITS DIAGONALS ARE EQUAL.

Proof. GIVEN PARALLELOGRAM ABCD, FIRST ASSUME EQUAL DIAGONALS, AC = BD.

1. $AB = CD$, $AD = BC$ (THM. 12-4)

2. $AC = BD$ (ASSUMED)

3. $\triangle ADC \cong \triangle BCD$ (SSS)

4. $\angle ADC = \angle BCD$ (CORR. PARTS)

5. ALSO $\angle ADC = \angle ABC$ AND $\angle BCD = \angle DAB$.   (THM. 12-2)

6. ALL FOUR ANGLES ARE EQUAL, SO ABCD IS A RECTANGLE.   (DEF. OF RECTANGLE)

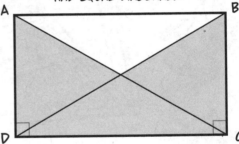

IF A PARALLELOGRAM'S DIAGONALS ARE EQUAL, SO ARE ADJACENT VERTEX ANGLES (SO THE DRAWING IS MISLEADING).

CONVERSELY, ASSUME ABCD IS A RECTANGLE.

A RECTANGLE OBVIOUSLY HAS EQUAL DIAGONALS.

1. $AD = BC$ (THM. 12-4)

2. $CD = CD$

3. $\angle ADC = \angle BCD = 90°$ (ASSUMED)

4. $\triangle ACD \cong \triangle BDC$ (SAS)

5. $AC = BD$ ∎   (CORR. PARTS)

TO SEE IF HIS PARALLELOGRAM BOOKSHELF IS RECTANGULAR, KEVIN NEEDS ONLY CHECK THAT ITS DIAGONALS ARE EQUAL.

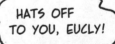

HATS OFF TO YOU, EUCLY!

THESE FLEXIBLE PARALLELOGRAMS SHOW THAT SIDE-SIDE-SIDE-SIDE DOESN'T WORK AS A CONGRUENCE TEST FOR QUADRILATERALS. TWO FOUR-SIDED FIGURES CAN HAVE ALL FOUR CORRESPONDING SIDES EQUAL BUT NOT BE CONGRUENT.

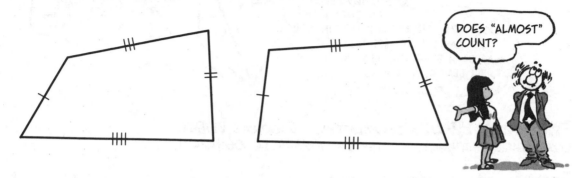

BUT TRIANGLES **DO** OBEY SIDE-SIDE-SIDE. A TRIANGLE WITH THREE GIVEN SIDES CAN HAVE ONLY ONE SHAPE. TRIANGLES, IN OTHER WORDS, ARE RIGID.

THIS EXPLAINS WHY THERE IS SO MUCH DIAGONAL BRACING IN THE WORLD. STRUTS AND TRUSSES ADD TRIANGLES TO STRUCTURES AND SO GIVE THEM "SHEAR STRENGTH," RESISTANCE TO FLEXING, DEFORMATION, AND COLLAPSE.

A SHEET OF PLYWOOD BACKING ALSO CREATES TRIANGES.

DIAGONALS CAN TELL US A LOT ABOUT QUADRILATERALS.

OH, MAN, DO I HAVE SOME STORIES ABOUT THIS ONE! TALK ABOUT **IRREGULAR!** AND NOT IN A GOOD WAY... JUST LAST NIGHT, IT WAS BLAH BLAH BLAH BLAH BLAH BLAH BLAH BLAH BLAH

HOW DO YOU MAKE IT SHUT UP?

Theorem 12-6. A QUADRILATERAL'S DIAGONALS **BISECT** EACH OTHER IFF THE QUADRILATERAL IS A **PARALLELOGRAM.**

Proof. LET THE DIAGONALS INTERSECT AT POINT E. FIRST WE SHOW THAT IF AE = EC AND DE = EB, THEN THE QUADRILATERAL IS A PARALLELOGRAM.

1. ∠1 = ∠2   (VERT. ∠s)

2. △AED ≅ △CEB   (SAS)

3. AD = BC   (CORR. PARTS)

4. SIMILARLY, △AEB ≅ △CED, SO AD = DC, AND BOTH PAIRS OF OPPOSITE SIDES ARE EQUAL.

5. ABCD IS A   (THM. 12-4)
   PARALLELOGRAM. ▌

△AED ≅ △CEB ⇒ AD = BC
△AEB ≅ △CED ⇒ AB = DC

CONVERSELY, ASSUME THAT ABCD IS A PARALLELOGRAM.

1. AD ∥ BC   (ASSUMED)

2. ∠3 = ∠6, ∠4 = ∠5   (PAR. POST.)

3. AD = BC   (THM. 12-4)

4. △AED ≅ △CEB   (ASA)

5. AE = EC, DE = EB;   (CORR. PARTS)
   THE DIAGONALS BISECT EACH OTHER. ▌

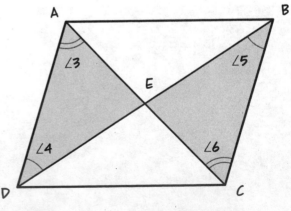

△AED ≅ △CEB ⇒ AE = EC, DE = EB

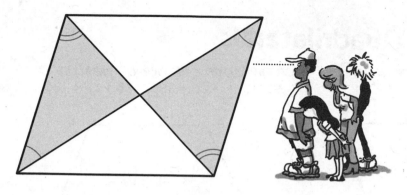

STARE AT THAT FIGURE A MINUTE AND ASK YOURSELF, WHAT IF THOSE FOUR CENTRAL ANGLES WERE **EQUAL?** WOULDN'T **ALL FOUR** TRIANGLES BE CONGRUENT? AND THEN WOULDN'T ALL FOUR SIDES BE **EQUAL?**

A QUADRILATERAL WITH FOUR EQUAL SIDES IS CALLED A **RHOMBUS,** AND WE CONCLUDE:

Corollary 12-6.1. A PARALLELOGRAM IS A RHOMBUS IFF ITS DIAGONALS ARE PERPENDICULAR.

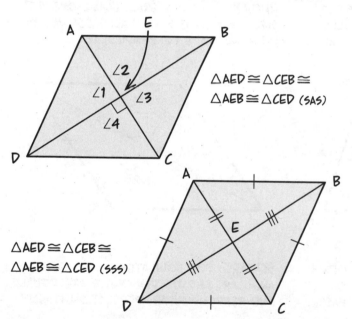

$\triangle AED \cong \triangle CEB \cong$
$\triangle AEB \cong \triangle CED$ (SAS)

$\triangle AED \cong \triangle CEB \cong$
$\triangle AEB \cong \triangle CED$ (SSS)

Proof. IF THE DIAGONALS ARE PERPENDICULAR, THEN

$$\angle 1 = \angle 2 = \angle 3 = \angle 4$$

AS ABCD IS A PARALLELOGRAM, AE = EC, DE = EB, AND ALL FOUR TRIANGLES ARE CONGRUENT BY SAS. HENCE AB = BC = DC = AD. ∎

CONVERSELY, IF

$$AB = BC = DC = AD$$

THEN THE FOUR TRIANGLES ARE CONGRUENT BY SSS. SO $\angle 1 = \angle 2 = \angle 3 = \angle 4$, SO THEY MUST ALL BE 90°. ∎

A RIGHT RHOMBUS IS A **SQUARE,** THE ONLY QUADRILATERAL WHOSE DIAGONALS ARE BOTH **EQUAL** AND EACH OTHER'S **PERPENDICULAR BISECTORS.**

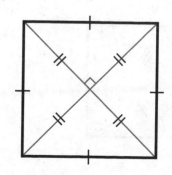

SO PERFECT! I'M JEALOUS!

A Gallery of Quadrilaterals

IN A GARDEN-VARIETY QUADRILATERAL, THE FOUR ANGLES ADD TO 360°.

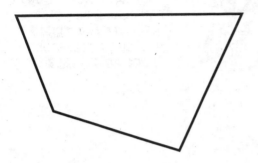

TRAPEZOID. ONE PAIR OF PARALLEL SIDES. ∠A + ∠C = 180°, ∠B + ∠D = 180°.

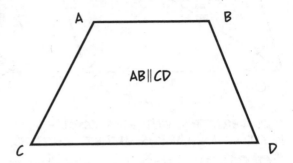

AB∥CD

PARALLELOGRAM. ALL OPPOSITE SIDES PARALLEL. OPPOSITE SIDES AND ANGLES EQUAL. DIAGONALS BISECT EACH OTHER.

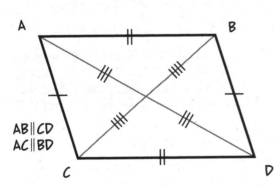

AB∥CD
AC∥BD

RHOMBUS. EQUILATERAL. EQUAL OPPOSITE SIDES IMPLIES IT'S A PARALLELOGRAM. ITS DIAGONALS ARE PERPENDICULAR.

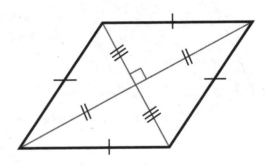

RECTANGLE. THE EQUIANGULAR QUADRILATERAL. OPPOSITE ANGLES ARE EQUAL, SO IT'S A PARALLELOGRAM—AND IT HAS EQUAL DIAGONALS.

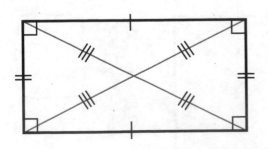

SQUARE. IT'S EQUILATERAL, IT'S EQUIANGULAR, AND ITS DIAGONALS ARE EQUAL AND PERPENDICULAR. PERFECTION ITSELF!

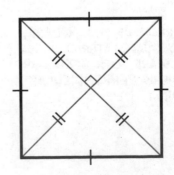

OF COURSE, SQUARE IS A PUT-DOWN TO SOME PEOPLE.

UNHIP, STRAITLACED, ORDINARY, NOT SO FUNNY, STODGY, AND DULL.

WHAT A GIGANTIC LOAD OF FLATTERY YOU HAVE, GRANDMA!

ON THE OTHER HAND, SQUARENESS IS, WELL, SQUARELY WITH US.

CARPENTERS USE **TRI-SQUARES**; CARTOONISTS USE **T-SQUARES**; FAIR NEGOTIATION IS CALLED **SQUARE DEALING** (EXCEPT IN NEW YORK REAL ESTATE, WHERE IT'S CALLED "STUPIDITY," I'M TOLD); AND WE HANG OUT IN **TOWN SQUARES**... SQUARES AREN'T SO BAD, AFTER ALL!

SQUARE MEAL

IN GEOMETRY, WE USE SQUARES IN **SQUARE CENTIMETERS, SQUARE INCHES, SQUARE FEET,** AND **SQUARE METERS.** WE ARE ABOUT TO EXPAND OUR IDEA OF MEASUREMENT BEYOND LENGTHS AND ANGLES AND INTO **TWO-DIMENSIONAL REGIONS** IN THE PLANE—USING **SQUARES.**

Exercises

1a. ∠1 = ?

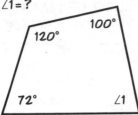

120°   100°

72°   ∠1

b. ∠1 + ∠2 = ?

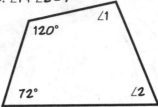

∠1

120°

72°   ∠2

c. ∠3 = ?

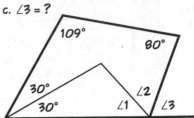

109°   80°

30°   ∠2

30°   ∠1   ∠3

2a. ARE THESE TWO PARALLELO-
GRAMS CONGRUENT (I.E., ALL
CORRESPONDING SIDES AND
ANGLES EQUAL)? WHY?

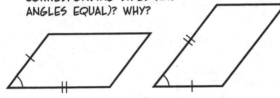

2b. ARE THESE TWO PARALLELOGRAMS
CONGRUENT? WHY?

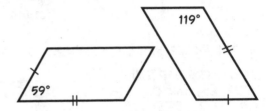

119°

59°

3. IN THIS FIGURE,
IF THE TWO
PARALLELOGRAMS
ARE CONGRUENT,
WHAT IS ∠1?

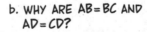

55°

∠1

4. IN QUADRILATERAL ABCD, THE
DIAGONALS ARE PERPENDICULAR,
AND BD BISECTS AC.

a. WHY IS △ABM ≅ △CBM?

b. WHY ARE AB = BC AND
AD = CD?

c. IS ∠BAD = ∠BCD? IS ∠ABC = ∠ADC?

THIS FIGURE IS CALLED A **KITE**.

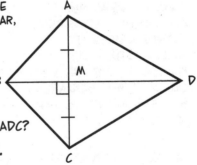

A

B   M   D

C

5a. IN THIS PINWHEEL-LOOKING
THING, FOUR RHOMBI SUR-
ROUND A SQUARE OF THE
SAME SIDE. WHAT ARE THE
ANGLES ∠1, ∠2, ∠3, ∠4?

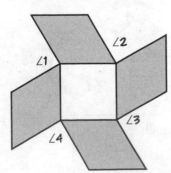

∠2

∠1

∠3

∠4

b. CONCLUDE THAT THIS PATTERN
FILLS, OR **TILES**, THE PLANE.

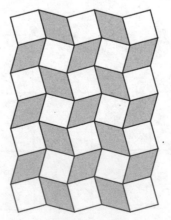

6. WOULD THIS WORK
FOR A RECTANGLE AND
A PARALLELOGRAM?

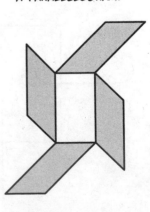

Chapter 13
AREA

WHEN PEOPLE TALK ABOUT THE SIZE OF THEIR PHONE OR TV, THEY BRAG ABOUT THEIR "SCREEN REAL ESTATE."

I BOUGHT THE HALF-ACRE MODEL.

REAL REAL ESTATE—LAND AND BUILDINGS—BRINGS US BACK TO THE ORIGINAL EGYPTIAN GEOMETRY PROBLEM: HOW DO YOU MEASURE AN AMOUNT OF LAND (OR A FLAT PIECE OF AN IMAGINARY PLANE, ANYWAY)?

A FARMER MIGHT SAY THAT REGIONS ARE THE SAME SIZE IF THEY GROW THE SAME AMOUNT OF CROPS, GIVEN THE SAME SOIL, WATER, AND CLIMATE. EACH OF THESE RECTANGULAR GARDENS GROWS TWELVE CABBAGES, DESPITE THE DIFFERENT DIMENSIONS.

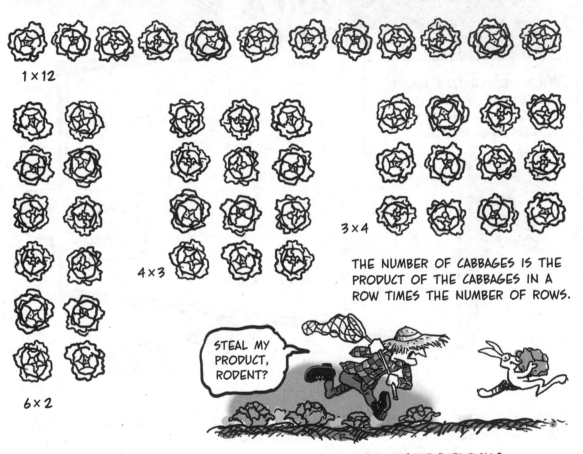

1 × 12

4 × 3

6 × 2

3 × 4

THE NUMBER OF CABBAGES IS THE PRODUCT OF THE CABBAGES IN A ROW TIMES THE NUMBER OF ROWS.

STEAL MY PRODUCT, RODENT?

THAT IS, A RECTANGLE'S AREA IS THE PRODUCT OF ITS HEIGHT (THE "VERTICAL" SIDE) TIMES ITS WIDTH (THE "HORIZONTAL" SIDE). NO SURPRISE THERE!

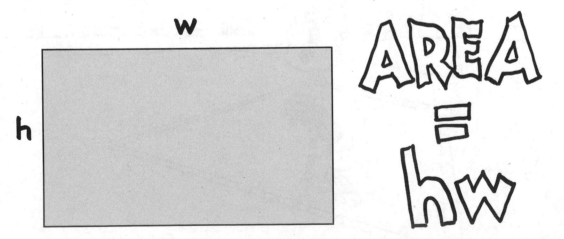

w

h

$$AREA = hw$$

A FARMER WOULD ALSO AGREE THAT NON-RECTANGULAR REGIONS HAVE AREAS, TOO.

JUST COUNT THE CABBAGES!

WE NEED TO BE A BIT CAREFUL HERE... SOME SHAPES WON'T HAVE AREAS, FOR INSTANCE, REGIONS THAT SPREAD TO INFINITY.

WHAT WOULD INFINITE CABBAGES LOOK LIKE?

VERY HARD TO WATER.

HOW DO WE FIND A FIGURE'S AREA? RECTANGLES ARE EASY. FOR MORE COMPLICATED SHAPES, WE GENERALLY SPLIT THEM INTO SIMPLER PIECES AND ADD UP THE SMALLER AREAS.

THIS IDEA OF ADDING UP THE PIECES IS KEY TO THE DEFINITION OF AREA.

Postulate 11.

EVERY "REASONABLE" REGION (SEE BELOW) OF THE PLANE HAS A MEASUREMENT CALLED ITS **AREA**. WRITING $\mathcal{A}(S)$ FOR THE AREA OF A REGION S, AREA HAS THE FOLLOWING PROPERTIES:

1. EVERY RECTANGLE HAS AN AREA EQUAL TO THE PRODUCT OF ITS TWO SIDES.

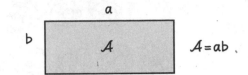

$\mathcal{A} = ab$.

2. IF A REGION IS MADE UP OF TWO OTHER REGIONS, ITS AREA IS THE SUM OF THEIRS.

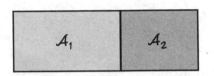

$\mathcal{A} = \mathcal{A}_1 + \mathcal{A}_2$

3. CONGRUENT REGIONS HAVE EQUAL AREAS.

$\mathcal{A}_3 = \mathcal{A}_4$

IN FARMING TERMS, IF A GARDEN IS DIVIDED INTO TWO PARTS, THE CABBAGES IN ONE PART PLUS THE CABBAGES IN THE OTHER PART MAKE THE TOTAL CABBAGES IN THE GARDEN—EVEN ALLOWING FOR SPLIT CABBAGES.

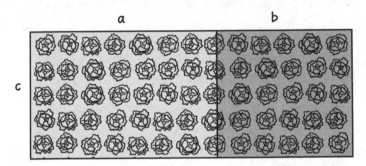

FOR RECTANGLES, THIS AMOUNTS TO THE **DISTRIBUTIVE LAW** THAT YOU LEARNED IN ALGEBRA:

$$(a + b)c = ac + bc$$

AND WHAT'S A "REASONABLE" REGION? THAT'S SOMETHING WE'LL LEAVE DELIBERATELY VAGUE FOR NOW. JUST THINK OF IT AS A SHAPE MADE BY GLUING TOGETHER A FINITE NUMBER OF TRIANGLES.

AND TRIANGLES DO HAVE AREAS!
LET'S START WITH RIGHT TRIANGLES.

Theorem 13-1. A RIGHT TRIANGLE'S AREA
IS HALF THE PRODUCT OF ITS LEGS. IF THE LEGS
HAVE LENGTHS a AND b, THEN THE AREA A IS

$$A = \frac{ab}{2}$$

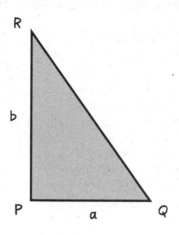

Proof. THIS IS BECAUSE A RIGHT TRIANGLE
IS HALF A RECTANGLE. LET'S SUPPOSE THAT
THE RIGHT TRIANGLE △PQR HAS PQ=a, PR=b.

1. DRAW A RECTANGLE PQRS
   WITH SIDES a AND b.

2. A(PQRS) = ab   (POST. 11.1)

3. △PQR ≅ △SRQ   (SAS/SSS)

4. A(△PQR) = A(△SRQ)   (POST. 11.3)

5. A(△PQR) + A(△SRQ) = A(PQRS)   (POST. 11.2)

6. A(△PRQ) + A(△SRQ) = ab   (SUBST.)

7. 2A(PQR) = ab   (SUBST.)

8. A(△PQR) = $\frac{ab}{2}$ ∎   (ALGEBRA)

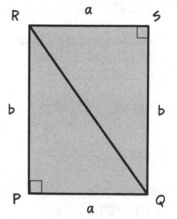

DON'T YOU THINK IT'S
EASIER JUST TO LOOK
AT THE PICTURES THAN
TO DECIPHER ALL THAT,
AHEM, STUFF?

OF COURSE!
BUT THEN, I'M A
CARTOONIST...

WE FIND A RANDOM TRIANGLE'S AREA BY DIVIDING IT INTO TWO RIGHT TRIANGLES. FIRST DROP A PERPENDICULAR FROM ANY VERTEX TO THE OPPOSITE SIDE.

Definition.

GIVEN ANY VERTEX OF A TRIANGLE, THE SIDE OPPOSITE THE VERTEX IS CALLED A **BASE,** AND THE PERPENDICULAR SEGMENT FROM VERTEX TO BASE IS AN **ALTITUDE** OF THE TRIANGLE.

HERE BC IS THE BASE, AND AD IS THE ALTITUDE.

ANY SIDE CAN BE A BASE!

Theorem 13-2. A TRIANGLE'S AREA IS HALF ITS BASE TIMES ITS ALTITUDE.

Proof. GIVEN △ABC, DRAW THE ALTITUDE AD. ∠ADB = ∠ADC = 90°. THEN, BY THEOREM 13-1,

$$\mathcal{A}(\triangle ABD) = \frac{(AD)(BD)}{2}$$

$$\mathcal{A}(\triangle ADC) = \frac{(AD)(DC)}{2}$$

BY THE AREA POSTULATE, THESE ADD UP TO THE TOTAL AREA:

$$\mathcal{A}(\triangle ABC) = \frac{(AD)(BD)}{2} + \frac{(AD)(DC)}{2}$$

$$= \frac{(AD)(BD + DC)}{2}$$

$$= \frac{(AD)(BC)}{2} \ \blacksquare$$

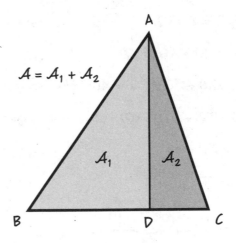

$$\mathcal{A} = \mathcal{A}_1 + \mathcal{A}_2$$

WE WILL OFTEN WRITE h FOR THE LENGTH OF THE ALTITUDE (h = HEIGHT!) AND b FOR THE LENGTH OF THE BASE. THEN THE TRIANGLE AREA FORMULA IS

$$\mathcal{A} = \frac{bh}{2}$$

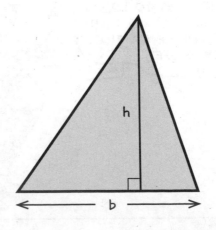

TO SEE HOW THIS
WORKS, WE ROLL
OUT THREE RECT-
ANGULAR SLICES OF
CHEESE, THREE FEET
BY FIVE FEET, AND
GIVE EVERYONE A
DULL KNIFE.

WE ASK THEM TO START CUTTING AT ONE CORNER, GO TO ANY POINT ON THE OP-
POSITE LONG SIDE, AND RETURN TO THE OTHER END OF THE FIRST LONG SIDE.

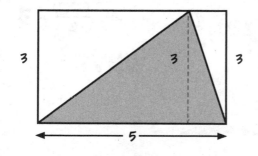

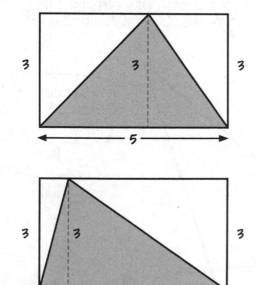

ALL TRIANGLES MADE THIS WAY HAVE
A BASE OF 5 AND AN ALTITUDE OF 3
(BECAUSE THE CHEESE IS RECTANGULAR).

SO ALL THREE TRIANGLES HAVE THE
SAME AREA!

$$A = \frac{bh}{2} = \frac{(5)(3)}{2}$$

$$= 7.5$$

HALF THE RECTANGLE.

TOO BAD
I'M LACTOSE
INTOLERANT...

STILL SLIGHTLY HUNGRY, BIANCA GETS ANOTHER CHEESE RECTANGLE AND CUTS A SMALLER TRIANGLE. THIS ONE HAS A 2-FOOT BASE AND AREA

$$A = \frac{(2)(3)}{2} = 3$$

JUST A SNACK!

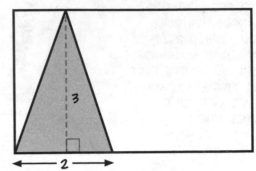

THEN SHE CUTS ANOTHER TRIANGLE, ALSO WITH A 2-FOOT BASE. THIS ONE SEEMS TO HAVE ITS ALTITUDE **OUTSIDE THE TRIANGLE.**

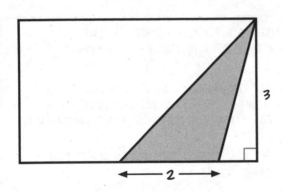

BUT EQUALLY FILLING!

THIS TRIANGLE'S AREA IS ALSO

$$A = \frac{(2)(3)}{2} = 3$$

WHY?

A TRIANGLE △ABC, WHERE THE PERPENDICULAR AD TO THE BASE LIES OUTSIDE THE TRIANGLE, IS THE **DIFFERENCE** BETWEEN TWO RIGHT TRIANGLES, △ADB AND △ACD.

$$A(\triangle ABC) + A(\triangle ACD) = A(\triangle ABD)$$

$$A(\triangle ABC) = A(\triangle ABD) - A(\triangle ACD)$$

$$= \frac{(h)(BD)}{2} - \frac{(h)(CD)}{2}$$

$$= \frac{(h)(BD - CD)}{2}$$

$$= \frac{(h)(BC)}{2}$$

A DELICIOUS MEAL **AND** FORMULA!

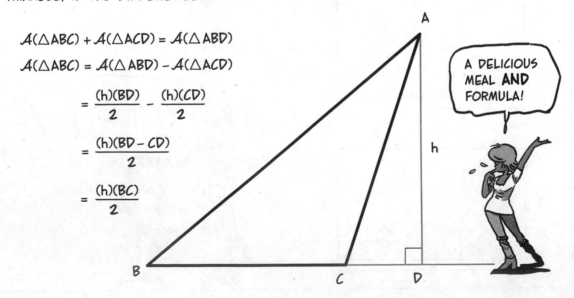

THINK OF A TRIANGLE'S ALTITUDE AS THE DISTANCE FROM A VERTEX TO THE **LINE CONTAINING THE BASE.** THE ALTITUDE MAY BE INSIDE OR OUTSIDE THE TRIANGLE.

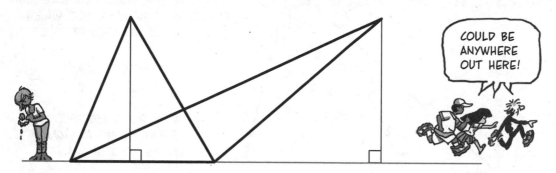

COULD BE ANYWHERE OUT HERE!

WE CAN SUM THESE THOUGHTS UP LIKE THIS:

Theorem 13-3. IF TWO TRIANGLES SHARE A BASE, AND EACH ONE HAS ITS THIRD VERTEX ON A LINE PARALLEL TO THE BASE, THEN THE TRIANGLES HAVE THE SAME AREA.

Proof. GIVEN $\triangle ABC$ AND $\triangle A'BC$ WITH $AA' \parallel BC$, THEN $\mathcal{A}(\triangle ABC) = \mathcal{A}(\triangle A'BC)$

1. DRAW ALTITUDES AD AND A'D'.   (CONST., P. 86)

2. $AD \perp \overline{BC}$, $A'D' \perp \overline{BC}$ (DEF. OF ALT.)

3. $AD \parallel A'D'$ (THM. 10-2)

4. $AA' \parallel DD'$ (ASSUMED)

5. AA'D'D IS A PARALLELOGRAM.   (DEF. OF PAR.)

6. $AD = A'D'$ (THM. 12-4)

7. THE TWO TRIANGLES HAVE EQUAL BASES AND ALTITUDES, SO THEIR AREAS ARE EQUAL. ∎

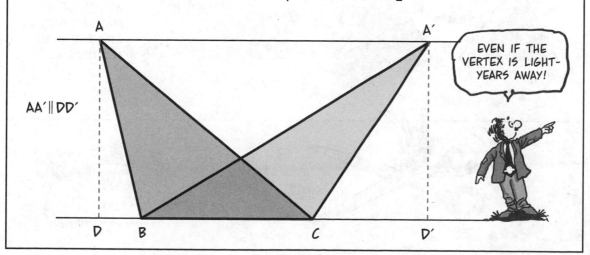

$AA' \parallel DD'$

EVEN IF THE VERTEX IS LIGHT-YEARS AWAY!

139

BUT WHICH SIDE IS "THE" BASE? AND WHICH SEGMENT IS "THE" ALTITUDE?

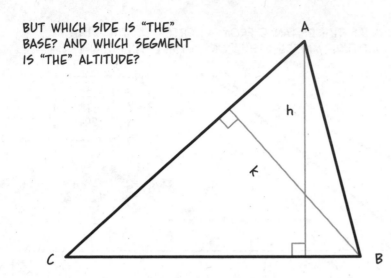

IN DERIVING THE FORMULA FOR A TRIANGLE'S AREA, WE SINGLED OUT ONE SIDE AS THE BASE. BUT WHICH SIDE? IT DOESN'T MATTER!

$$\frac{h(BC)}{2} = \frac{k(AC)}{2}$$

WHY DO THESE TWO EXPRESSIONS MAGICALLY COME TO THE SAME NUMBER?

HERE'S A CUT-AND-PASTE EXPLANATION: WE SPLIT THE TRIANGLE ALONG EACH ALTITUDE AND COPY THE PARTS TO FORM TWO RECTANGLES, ONE $h \times a$, THE OTHER $k \times b$. THE RECTANGLES CAN THEN BE REARRANGED INTO CONGRUENT FIGURES.

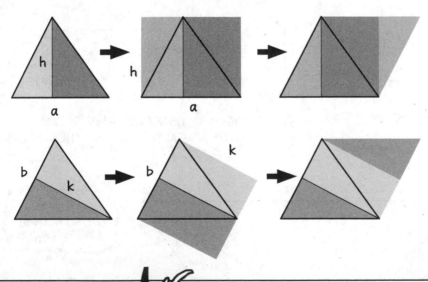

RESULT:

$$bk = ah$$

SO OF COURSE

$$\frac{bk}{2} = \frac{ah}{2}$$

THE OUTCOME IS THE SAME WITH BASE a AND ALTITUDE h, OR BASE b AND ALTITUDE k.

OKAY! I BELIEVE IT!

Areas of Quadrilaterals

WHAT'S THE AREA OF A RANDOM QUADRILATERAL? THERE IS A FIENDISH FORMULA BASED ON THE SIDES AND ANGLES, BUT WE ARE **NOT** GOING TO GET INTO IT.

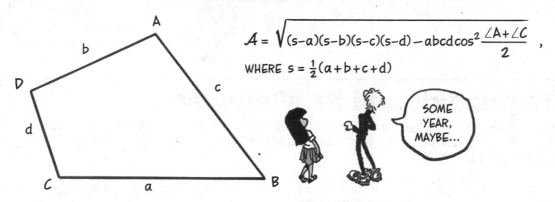

$$\mathcal{A} = \sqrt{(s-a)(s-b)(s-c)(s-d) - abcd\cos^2\frac{\angle A + \angle C}{2}},$$

WHERE $s = \frac{1}{2}(a+b+c+d)$

SOME YEAR, MAYBE...

OF COURSE, WE CAN ALWAYS DIVIDE A QUADRILATERAL INTO TWO TRIANGLES, FIND THEIR AREAS, AND ADD THOSE TOGETHER.

SCISSOR-HOOD IS POWERFUL!

LUCKILY, THE SPECIAL QUADRILATERALS WE SAW IN CHAPTER 12 HAVE AREAS THAT ARE EASY TO FIND.

AS SOON AS A QUADRI-
LATERAL HAS ONE PAIR OF
PARALLEL SIDES, IT ALSO
HAS AN **ALTITUDE.** THIS IS
THE DISTANCE BETWEEN
THE PARALLEL LINES.

THE LENGTH OF
ANY PERPENDIC-
ULAR SEGMENT
BETWEEN THEM.

Trapezoids and Parallelograms

IN A TRAPEZOID, A DIAGONAL CUTS THE TRAPEZOID INTO TWO TRIANGLES, EACH WITH
THE **SAME ALTITUDE.** (THE BASES ARE PARALLEL, SO THE ALTITUDE SEGMENT IS
PERPENDICULAR TO BOTH.)

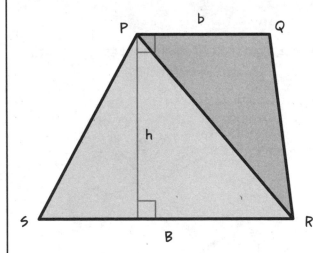

IF THE PARALLEL SIDES HAVE
LENGTH b AND B, AND THE
ALTITUDE IS h, THEN THE
TRAPEZOID'S AREA IS THE SUM
OF THE TRIANGLES' AREAS:

$$\mathcal{A}(PQRS) = \mathcal{A}(\triangle PQR) + \mathcal{A}(\triangle PRS)$$

$$= \frac{hb}{2} + \frac{hB}{2}$$

$$= h\frac{B+b}{2}$$

THE HEIGHT TIMES THE
AVERAGE OF THE BASES.

A PARALLELOGRAM'S FORMULA IS EVEN SIMPLER.

Theorem 13-4. A PARALLELOGRAM
WITH BASE b AND ALTITUDE h HAS AREA **hb.**

Proof. A PARALLELOGRAM
IS JUST A SPECIAL KIND OF
TRAPEZOID, SO:

$$\mathcal{A} = h\frac{b+b}{2}$$

$$= h\frac{2b}{2}$$

$$= hb \ \blacksquare$$

HMM...
SOUNDS
FAMILIAR...

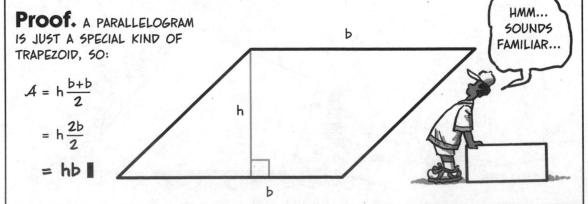

A PARALLELOGRAM'S AREA FORMULA LOOKS LIKE A RECTANGLE'S: BASE × ALTITUDE. THE ONLY DIFFERENCE IS THAT A RECTANGLE'S ALTITUDE EQUALS THE LENGTH OF ONE SIDE.

SAME AREA!

h

h

AS A MATTER OF FACT, A SURGEON CAN TURN ANY PARALLELOGRAM INTO A RECTANGLE BY SNIPPING OFF A RIGHT TRIANGLE FROM ONE END AND GRAFTING IT TO THE OTHER.

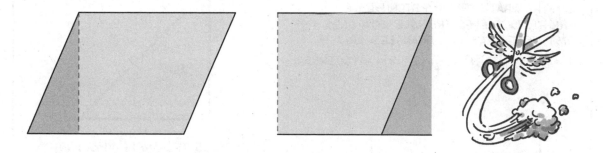

AS WITH TRIANGLES, TWO PARALLELOGRAMS WITH THE SAME BASE HAVE THE SAME AREA IF THE OPPOSITE SIDES LIE ON THE SAME LINE (NECESSARILY PARALLEL TO THE BASE, BECAUSE THESE ARE PARALLELOGRAMS!). THAT FORMULA AGAIN?

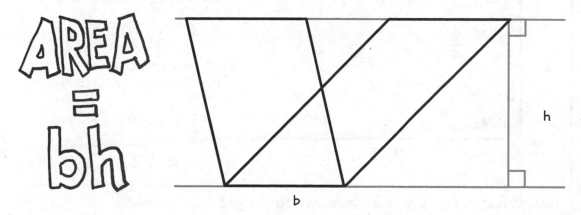

AREA = bh

b

h

A CONCRETE EXAMPLE—WITH REAL CONCRETE—IS THIS DAM BUILT BETWEEN TWO SLOPES. GIVEN ITS MEASUREMENTS, WE CAN FIND THE AREA OF ITS TRAPEZOIDAL FACE.*

$A = \frac{1}{2}h(B+b)$

$\quad = \frac{1}{2}(32m)(45m + 28m)$

$\quad = (16m)(73m)$

$\quad = \mathbf{1,168m^2}$

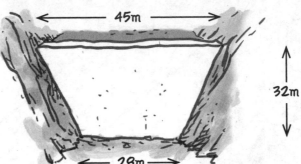

THE UNIT OF AREA HERE IS METERS × METERS, OR **SQUARE METERS, m²**. IF THE LENGTHS WERE MEASURED IN SOME OTHER UNITS, LIKE INCHES, FEET, ETC., THE UNIT OF AREA WOULD BE SQUARE INCHES, SQUARE FEET, ETC.

ONE SQUARE UNIT IS THE AREA OF A SQUARE ONE UNIT ON A SIDE.

NOW LET'S USE AN AREA TO CALCULATE A LENGTH: WHAT'S THE HYPOTENUSE OF A RIGHT, ISOSCELES TRIANGLE WITH LEGS = 1? FIRST NOTE THAT THE TRIANGLE'S AREA IS $A = \frac{1}{2}bh = \frac{1}{2}(1)(1) = \frac{1}{2}$. LET x = HYPOTENUSE.

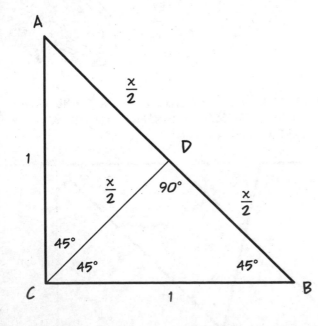

SUPPOSE THE ANGLE BISECTOR AT C INTERSECTS AB AT D.

$CD \perp AB$ AND $DB = \frac{x}{2}$ (WHY?)

ALSO $CD = \frac{x}{2}$ (WHY?), SO

$$A = \left(\frac{x}{2}\right)\left(\frac{x}{2}\right) = \frac{x^2}{4}$$

THE TWO EXPRESSIONS FOR THE TRIANGLE'S AREA ARE EQUAL:

$\frac{x^2}{4} = \frac{1}{2}$, SO $x^2 = 2$ AND

$$x = \sqrt{2}$$

*EXCEPT THAT IN REAL LIFE, DAMS ARE CURVED TO INCREASE THEIR STRENGTH.

ALL THE FORMULAS IN THIS CHAPTER FOR FINDING AREAS DEPEND ON THE REGION'S **ALTITUDE.**

NOW WE MIGHT WONDER, CAN WE FIND THIS ALTITUDE'S SIZE FROM THE DIMENSIONS OF A TRIANGLE?

WE CAN ALWAYS **CONSTRUCT** AN ALTITUDE AS ON PAGE 86: IT'S THE PERPENDICULAR TO A LINE FROM A POINT OFF THE LINE.

BUT WHAT IS ITS **LENGTH?** HOW DO WE KNOW WHAT **THAT** IS? THE ANSWER IS: **WE DON'T,** EXCEPT BY MEASURING IT.

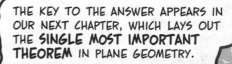

AT THIS POINT, MUCH REMAINS MYSTERIOUS ABOUT TRIANGLES. FOR INSTANCE, GIVEN TWO SIDES AND THE ANGLE BETWEEN THEM, SHOULDN'T WE BE ABLE TO SOLVE FOR THE REST? BUT **HOW?**

THE KEY TO THE ANSWER APPEARS IN OUR NEXT CHAPTER, WHICH LAYS OUT THE **SINGLE MOST IMPORTANT THEOREM** IN PLANE GEOMETRY.

145

Exercises

1. FIND THE AREAS.

a.

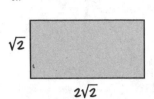

$\sqrt{2}$

$2\sqrt{2}$

b.

5

3

c.

12

5

d.

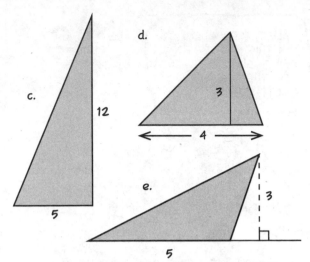

3

4

e.

3

5

2a. IF THE AREA OF PARALLELO-GRAM ABCD IS 22.548 AND AE = 5.637, WHAT IS DC?

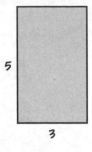

A   B

D   E   C

P

R   S   Q

2b. IN ISOSCELES △PQR, IF PS=6 AND RS=1, WHAT IS THE AREA $\mathcal{A}(\triangle PQR)$?

3. IF THE AREA OF THE RECTANGLE ON THE LEFT IS 4,676,004, WHAT IS THE AREA OF THE PARALLELOGRAM ON THE RIGHT?

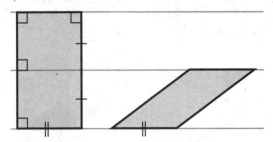

4a. IF AP IS THE MEDIAN FROM A (I.E. BP=PC), SHOW THAT

$$\mathcal{A}(\triangle ABP)=\mathcal{A}(\triangle APC)$$

4b. IF M IS A POINT ON THE MEDIAN, SHOW THAT

$$\mathcal{A}(\triangle AMB)=\mathcal{A}(\triangle AMC)$$

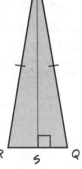

A

M

B   P   C

4c. IF THE MEDIAN CQ INTERSECTS AP AT M, SHOW THAT

$$\mathcal{A}(\triangle AMC) = \mathcal{A}(\triangle BMC)$$
$$= \mathcal{A}(\triangle AMB)$$

DO YOU THINK THAT ALL THREE MEDIANS PASS THROUGH M?

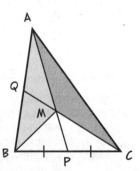

A

Q

M

B   P   C

5. A PARALLELOGRAM'S DIAGONALS INTERSECT AT P. SHOW THAT ANY LINE THROUGH P DIVIDES THE PARAL-LELOGRAM INTO TWO EQUAL AREAS. (HINT: FIND CONGRUENCES.)

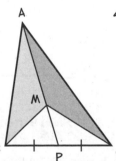

P

6. THIS FIGURE SUGGESTS WHAT VALUE FOR THIS ENDLESS SUM?

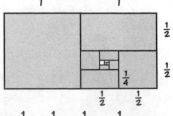

1   1

1    $\frac{1}{2}$

$\frac{1}{4}$ $\frac{1}{2}$

$\frac{1}{2}$ $\frac{1}{2}$

$$1 + \frac{1}{2} + \frac{1}{4} + \frac{1}{8} + \frac{1}{16} + \frac{1}{32} + \frac{1}{64} + \frac{1}{128} + \dots$$

Chapter 14
THE PYTHAGOREAN THEOREM

KEVIN HAS BEEN VISITING HIS FRIEND ACE. KEVIN LEAVES, WALKS FOUR BLOCKS SOUTH, TURNS WEST PAST COCO'S CONDO, AND STOPS THREE BLOCKS LATER, AT BETTY'S BARBECUE. A CROW MEANWHILE TAKES THE DIRECT ROUTE.

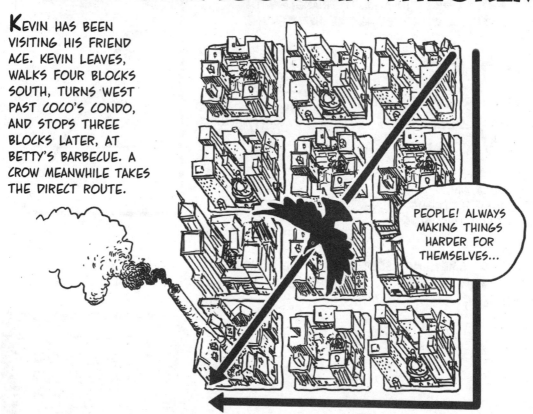

WHAT'S THE DISTANCE "AS THE CROW FLIES"? HOW LONG WOULD THE DIRECT ROUTE FROM ACE'S TO BETTY'S BE IF ALL THOSE BUILDINGS WEREN'T IN THE WAY? GIVEN A RIGHT TRIANGLE WITH KNOWN LEGS, WHAT IS ITS **HYPOTENUSE**?

THE ANSWER IS SIMPLE, SURPRISING, AND JUSTLY FAMOUS: **THE PYTHAGOREAN THEOREM.**

Theorem 14-1. IN A RIGHT TRIANGLE WITH SIDES a, b, AND c, WITH c BEING THE HYPOTENUSE,

$$a^2 + b^2 = c^2$$

OR, TAKING THE SQUARE ROOT,

$$c = \sqrt{a^2 + b^2}$$

THIS FORMULA GIVES THE **HYPOTENUSE** IN TERMS OF THE OTHER TWO **SIDES.**

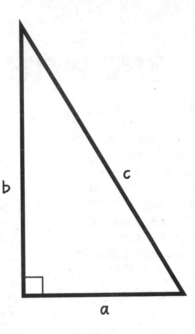

IT ALSO SAYS SOMETHING ABOUT **AREAS.** THE NUMBERS a^2, b^2, AND c^2 ARE THE AREAS OF SQUARES WITH SIDES a, b, AND c, RESPECTIVELY: THE **SQUARES ON THE SIDES OF THE TRIANGLE.**

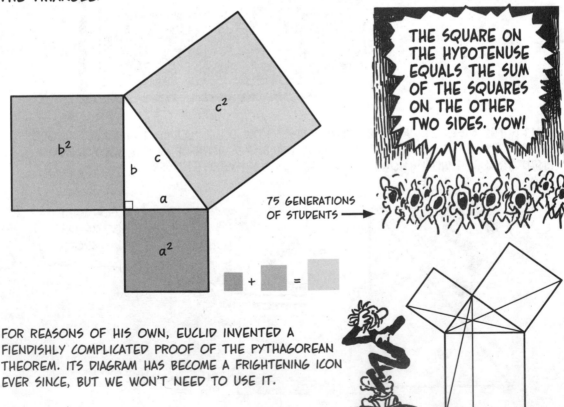

THE SQUARE ON THE HYPOTENUSE EQUALS THE SUM OF THE SQUARES ON THE OTHER TWO SIDES. YOW!

75 GENERATIONS OF STUDENTS →

FOR REASONS OF HIS OWN, EUCLID INVENTED A FIENDISHLY COMPLICATED PROOF OF THE PYTHAGOREAN THEOREM. ITS DIAGRAM HAS BECOME A FRIGHTENING ICON EVER SINCE, BUT WE WON'T NEED TO USE IT.

THIS ELEGANT FORMULA HAS INSPIRED MATHEMATICIANS ALL OVER THE WORLD (AND BEYOND? WHO KNOWS?) TO CREATE MORE THAN **ONE HUNDRED PROOFS.** IN THIS BOOK, WE OFFER ONLY FOUR(!). THE FIRST, BASED ON AREA, WAS KNOWN IN CENTRAL ASIA AND CHINA, THOUGH IT'S HARD TO SAY WHERE IT FIRST POPPED UP.

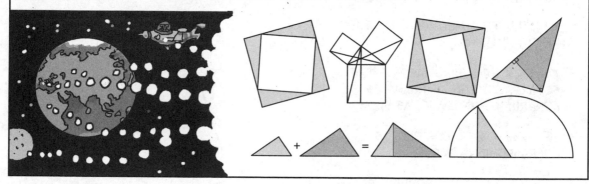

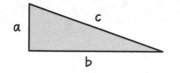

The Basic Idea of Proof #1:

STARTING WITH A RIGHT TRIANGLE, ARRANGE FOUR COPIES OF IT IN TWO DIFFERENT WAYS, AS SHOWN AT LEFT.

THESE TWO LARGE SQUARES HAVE THE SAME AREA, SINCE BOTH HAVE SIDE a + b, SO

$$a^2 + b^2 + \text{FOUR TRIANGLES}$$

$$= c^2 + \text{FOUR TRIANGLES}$$

SUBTRACTING THE FOUR TRIANGLES GIVES THE RESULT.

$$a^2 + b^2 = c^2$$

THE WHITE REGIONS ARE EQUAL.

A PROPER PROOF, WHICH FILLS IN THE DETAILS, APPEARS ON THE NEXT PAGE.

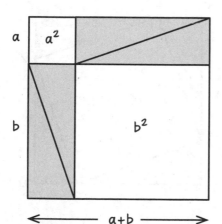

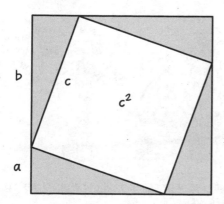

Proof.

WE ARE GIVEN A RIGHT TRIANGLE △ABC WITH SIDES a, b, AND c. ∠C IS THE RIGHT ANGLE, c THE HYPOTENUSE.* ITS AREA, $\mathcal{A}(\triangle ABC)$, IS ab/2.

1. CONSTRUCT SQUARE ABDE ON THE HYPOTENUSE AB.

2. ON EACH REMAINING SIDE OF THE SQUARE, CONSTRUCT TRIANGLES CONGRUENT TO △ABC, AS SHOWN.

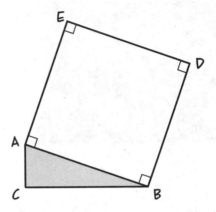

3. ∠F =∠G =∠H =∠C = 90°, AS CORRESPONDING PARTS.

4. ∠1 = ∠4, AS CORRESPONDING PARTS.

5. ∠1 +∠2 = 90° (ACUTE ANGLES IN A RIGHT TRIANGLE ARE COMPLEMENTARY).

6. ∠2 +∠4 = 90° (SUBSTITUTION).

7. ∠3 = 90° (ABDE IS A SQUARE).

8. ∠2 +∠3 +∠4 = 180°

9. F-A-C, THAT IS, FAC IS A LINE SEGMENT.

10. FC = a + b (RULER POSTULATE).

11. SIMILARLY, CH, HG, AND FG ARE SEGMENTS OF LENGTH a + b, SO CFGH IS A SQUARE OF SIDE a + b.

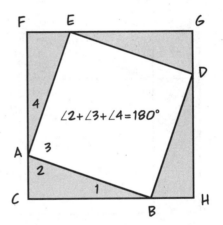

$\angle 2 + \angle 3 + \angle 4 = 180°$

12. $\mathcal{A}(CFGH) = \mathcal{A}(ABDE) + 4\,\mathcal{A}(\triangle ABC)$ (AREA POSTULATE), SO

$$(a + b)^2 = c^2 + 4\left(\frac{ab}{2}\right)$$

$$a^2 + 2ab + b^2 = c^2 + 2ab$$

$$a^2 + b^2 = c^2 \blacksquare$$

CON-VINCED?

SURE, BUT SO WHAT?

*WE GENERALLY LABEL A TRIANGLE'S SIDE WITH THE LOWERCASE LETTER CORRESPONDING TO THE OPPOSITE VERTEX'S CAPITAL LETTER. SO SIDE c IS OPPOSITE VERTEX C, ETC.

FOR STARTERS, WE CAN ANSWER THE ORIGINAL QUESTION: WHAT'S THE HYPOTENUSE OF A RIGHT TRIANGLE WITH LEGS 3 AND 4? HOW FAR DID THE CROW FLY?

NICE TO SEE YOU! A REAL RELIEF, IN FACT...

BY PYTHAGORAS,

$$c^2 = a^2 + b^2$$
$$= 3^2 + 4^2 = 9 + 16$$
$$= 25$$
$$c = \sqrt{25}$$
$$= 5$$

THIS IS ONE OF THE RARE RIGHT TRIANGLES HAVING ALL THREE SIDES EQUAL TO WHOLE NUMBERS. A FEW OTHER TRIPLES OF RIGHT-TRIANGLE INTEGER SIDES:

5, 12, 13
7, 24, 25
8, 15, 17
9, 40, 41

FOR INSTANCE:

$$8^2 + 15^2 = 64 + 225 = 289$$
$$17^2 = 289$$

17

8

15

LATER, WE'LL SEE HOW TO FIND THESE MYSTERIOUS "PYTHAGOREAN TRIPLES."

MOSTLY, THOUGH, WHOLE-NUMBER LEGS GO WITH AN IRRATIONAL HYPOTENUSE.

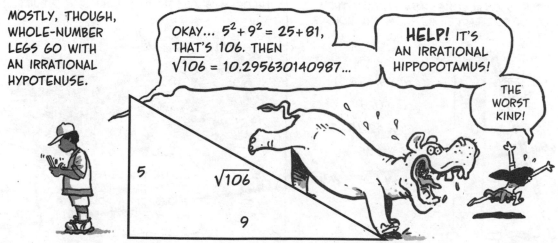

OKAY... $5^2 + 9^2 = 25 + 81$, THAT'S 106. THEN $\sqrt{106} = 10.295630140987...$

HELP! IT'S AN IRRATIONAL HIPPOPOTAMUS!

THE WORST KIND!

5

$\sqrt{106}$

9

Pythagoras in Real Life

MOMO'S FAMILY IS DOING WORK ON TOP OF THEIR HOUSE, INCLUDING RESHINGLING THE ROOF AND REPLACING A SKYLIGHT.

THIS LEADS TO SEVERAL PYTHAGOREAN PROBLEMS.

GLAD TO HELP!

GIVEN THESE DIMENSIONS IN METERS, WHAT IS THE ROOF'S AREA? IT CONSISTS OF TWO RECTANGLES. IGNORE THE SKYLIGHT.

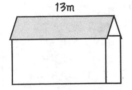 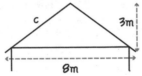

$$c^2 = 3^2 + (\tfrac{1}{2} \cdot 8)^2 = 9 + 16 = 25$$
$$c = \sqrt{25} = 5m$$

AREA OF ROOF:

$$2c(13) = (10m)(13m) = \mathbf{130\,m^2}$$

THE SKYLIGHT LOOKS LIKE THIS. HOW BIG A SHEET OF GLASS IS NEEDED? (DIMENSIONS ARE IN METERS.)

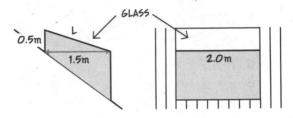

$$L^2 = (0.5)^2 + (1.5)^2 = 2.5$$
$$L = \sqrt{2.5} \approx 1.58m$$

THE GLASS IS ABOUT $2m \times 1.58m$.

THE GUTTER IS 6m OFF THE GROUND; THE LADDER IS 6.5m LONG. HOW FAR CAN THE FOOT OF THE LADDER BE FROM THE HOUSE?

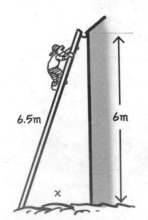

$$x^2 + 6^2 = (6.5)^2$$
$$x^2 = (6.5)^2 - 6^2$$
$$= 42.25 - 36$$
$$= 6.25$$
$$x = \sqrt{6.25} = \mathbf{2.5m}$$

A BUILDER IGNORANT OF THE PYTHAGOREAN THEOREM IS CALLED AN **UNEMPLOYED** BUILDER.

I LOVE THE DRIPS, THE DRAFTS, THE GLASS, THE DOORS, THE WHOLE EDGY POST-COMFORT AESTHETIC!

Some Special Triangles

IN THE LAST CHAPTER, WE WORKED HARD TO FIND THE HYPOTENUSE OF A 45°-45° ISOSCELES RIGHT TRIANGLE. NOW IT'S A SNAP.

$$c^2 = 1^2 + 1^2$$
$$= 1 + 1 = 2$$
$$c = \sqrt{2}$$

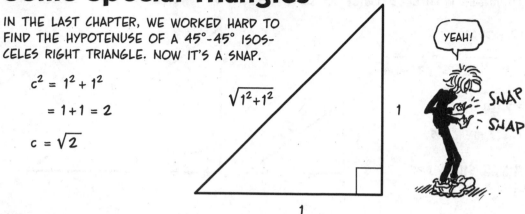

ANOTHER NICE ONE IS THE 30°-60° RIGHT TRIANGLE. ITS SHORTEST SIDE IS HALF THE HYPOTENUSE (SEE P. 112), AND NOW WE CAN FIND THE OTHER LEG. LET s BE THE SHORT LEG, 2s THE HYPOTENUSE, AND x THE UNKNOWN, LONGER LEG. THEN:

$$s^2 + x^2 = (2s)^2$$
$$x^2 = 4s^2 - s^2 = 3s^2$$
$$x = s\sqrt{3}$$

THE LONG LEG IS $\sqrt{3}$ TIMES THE SHORT LEG.

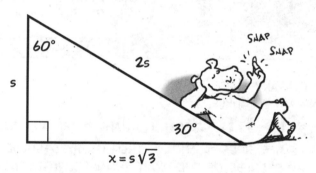

AN EQUILATERAL TRIANGLE'S ALTITUDE DIVIDES THE TRIANGLE INTO TWO 30°-60°-90° TRIANGLES. IF THE SIDE IS s, THE EQUILATERAL TRIANGLE'S **ALTITUDE** h IS $(s/2)\sqrt{3}$.

$$h = \frac{s\sqrt{3}}{2}, \quad b = s$$

ITS **AREA** IS

$$\frac{s}{2}\left(\frac{s\sqrt{3}}{2}\right)$$
$$= \frac{s^2\sqrt{3}}{4}$$

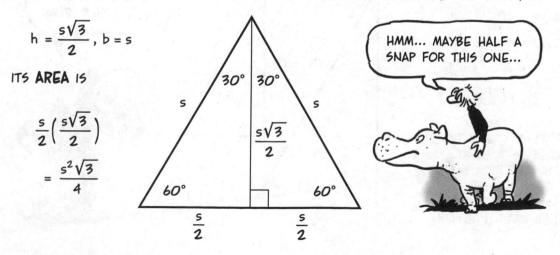

LET'S RETURN TO THE 3-4-5 TRIANGLE FROM KEVIN'S WALK AND DRAW ANOTHER RIGHT TRIANGLE WITH LEGS **TWICE AS LONG** (6 AND 8). WHAT IS ITS HYPOTENUSE?

BY PYTHAGORAS,

$$x = \sqrt{6^2 + 8^2} = \sqrt{36 + 64}$$

$$= \sqrt{100}$$

$$= \mathbf{10}, \text{ TWICE 5}$$

ALL THREE SIDES ARE DOUBLED.

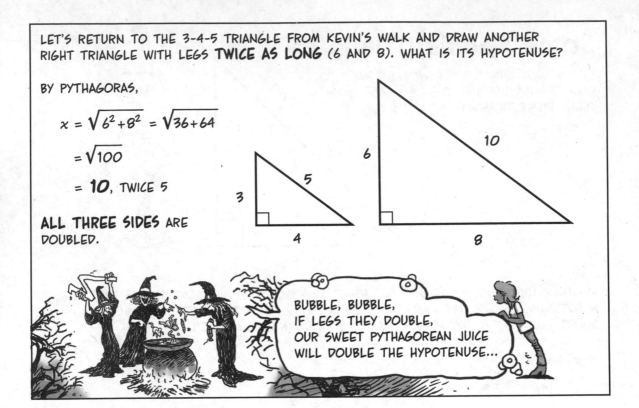

BUBBLE, BUBBLE,
IF LEGS THEY DOUBLE,
OUR SWEET PYTHAGOREAN JUICE
WILL DOUBLE THE HYPOTENUSE...

THIS WORKS FOR ANY RIGHT TRIANGLE AND ANY MULTIPLE.

Corollary 14-1.1. SUPPOSE r, a, AND b ARE ANY POSITIVE NUMBERS. IF A RIGHT TRIANGLE HAS SIDES ra AND rb, THEN ITS HYPOTENUSE IS $r\sqrt{a^2 + b^2}$, THAT IS, r TIMES THE HYPOTENUSE OF THE RIGHT TRIANGLE WITH LEGS a AND b.

Proof. PURE ALGEBRA.

$$c^2 = (ra)^2 + (rb)^2$$

$$= r^2 a^2 + r^2 b^2$$

$$= r^2 (a^2 + b^2)$$

$$c = r(\sqrt{a^2 + b^2})$$

IF WE MULTIPLY TWO SIDES OF A RIGHT TRIANGLE BY A FACTOR, THE THIRD SIDE IS ALSO MULTIPLIED BY THAT **SAME FACTOR**.

HERE (GLUG) IS A (GLUG) NEW IDEA!!

UP TO NOW, WE HAVE LOOKED AT **CONGRUENCE,** WHEN TWO FIGURES HAVE THE SAME SHAPE **AND** SIZE...

AND AT **AREA,** WHEN TWO FIGURES MAY HAVE THE SAME **SIZE** BUT DIFFERENT **SHAPES.**

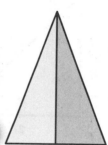

NOW WE'VE STUMBLED ON A SITUATION WHERE TWO FIGURES HAVE THE SAME **SHAPE** BUT DIFFERENT **SIZES.**

ONE FIGURE IS A **SCALED-UP VERSION** OF THE OTHER ONE...

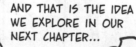

IT'S LIKE A PHOTOGRAPHIC ENLARGEMENT, A BLOWUP!

AND THAT IS THE IDEA WE EXPLORE IN OUR NEXT CHAPTER...

HEY, SMILE?

Exercises

1. IN THE FOLLOWING RIGHT TRIANGLES, FIND THE UNKNOWN SIDE.

a.
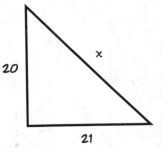
20
x
21

b.
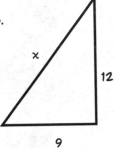
x
12
9

c.
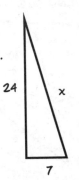
24
x
7

d.
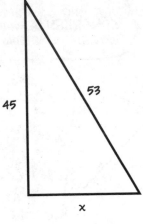
45
53
x

2. FIND THE UNKNOWN SIDE.

a.
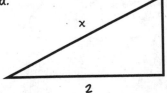
x
1
2

b.
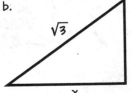
$\sqrt{3}$
1
x

c.
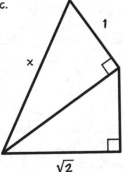
1
x
1
$\sqrt{2}$

3. IN EUCLID'S INFAMOUS DIAGRAM, SHOW THAT

$\triangle ADC \cong \triangle BDE$

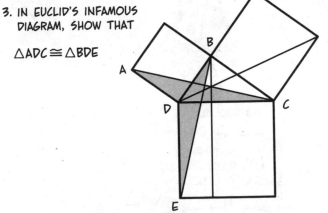

4. WHAT IS THE AREA OF THIS TRAPEZOID?

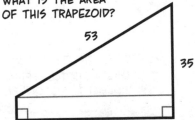
53
35
7

5. THESE TWO PARALLELOGRAMS HAVE TOP AND BOTTOM SIDES ON THE SAME PARALLEL LINES. THEY ALSO BOTH HAVE AREA = 32. WHAT'S THE LENGTH OF THE SIDE MARKED r?

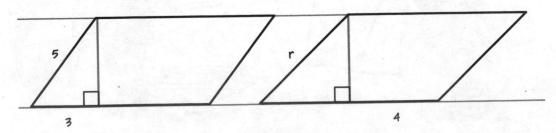
5
3
r
4

Chapter 15
SIMILARITY

AFTER ADDING, SUBTRACTING, AND MULTIPLYING SEGMENTS, WHAT'S LEFT?

WHAT DOES IT MEAN THAT TWO THINGS HAVE THE SAME SHAPE BUT DIFFERENT SIZES? LET'S THINK ABOUT PHOTO-GRAPHIC ENLARGEMENTS.

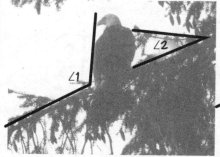

WARRRK!

IN AN ENLARGEMENT, ALL **ANGLES** ARE THE SAME AS IN THE ORIGINAL PHOTO. ANY TWO LINES IN ONE IMAGE MEET AT THE SAME ANGLE IN THE OTHER.

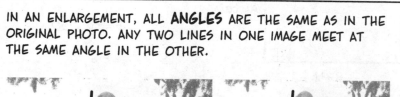

∠1 = ∠1′, ∠2 = ∠2′

OTHERWISE, SOME-BODY MESSED UP AT THE PHOTO LAB!

ALSO:

IN AN ENLARGEMENT, HORIZONTAL AND VERTICAL DISTANCES SCALE UP **EQUALLY**. A TWO-TIMES (2×) ENLARGEMENT DOUBLES HEIGHTS **AND** WIDTHS EVERYWHERE IN THE IMAGE.

BY COROLLARY 14-1.1, THIS IMPLIES THAT **ALL** LENGTHS ARE DOUBLED. ANY SEGMENT IN THE ORIGINAL IS THE HYPOTENUSE OF A RIGHT TRIANGLE WITH HORIZONTAL AND VERTICAL SIDES, WHICH DOUBLE IN THE ENLARGEMENT.

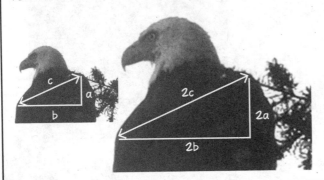

THIS IS CLEARLY TRUE OF ANY SCALING FACTOR, NOT JUST 2. A 3× ENLARGEMENT TRIPLES EVERY LENGTH.

YOU'RE NOT SO SPECIAL!!

SCALING A RECTANGLE BY A POSITIVE FACTOR t MAKES A NEW RECTANGLE WITH SIDES t TIMES THE SIDES OF THE ORIGINAL.

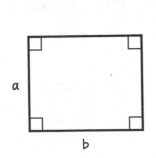

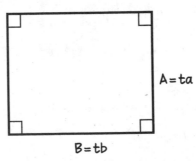

$A = ta$

$B = tb$

IF THE ORIGINAL SIDES ARE a AND b, THEN THE SCALED RECTANGLE HAS SIDES A AND B, WHERE

$$A = ta, \quad B = tb \quad \text{SO}$$

$$t = \frac{A}{a}, \quad t = \frac{B}{b}$$

THAT IS,

WE'LL SPEND THE **WHOLE CHAPTER** WITH THIS EQUATION!

$$\frac{A}{a} = \frac{B}{b}$$

ANY EQUALITY BETWEEN TWO FRACTIONS OR RATIOS IS CALLED A

PROPORTION.

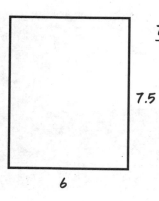

$$\frac{7.5}{5} = \frac{6}{4} \quad (= 1.5)$$

7.5

6

5

4

WE SAY THAT THE NUMERATORS (HERE 7.5 AND 6) AND THE DENOMINATORS (5 AND 4— IN THAT ORDER) ARE **PROPORTIONAL** OR **IN PROPORTION** TO EACH OTHER. "7.5 IS TO 5 AS 6 IS TO 4."

SEGMENTS DIVIDING SEGMENTS? WAIT... WHAT?

ADMITTEDLY, DIVIDING SEGMENTS IS MORE CHALLENGING THAN ADDING THEM (LAYING THEM END TO END) OR MULTIPLYING THEM (MAKING THEM THE SIDES OF A RECTANGLE AND FINDING THE AREA).

BUT $\frac{PQ}{QR}$?

P   Q   R

PR = PQ + QR

S   T

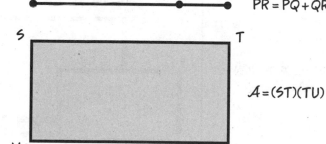

$\mathcal{A} = (ST)(TU)$

V   U

TO GET COMFORTABLE WITH THIS IDEA, LET'S GREASE THE GEARS OF AN EQUATION MUNCHER AND PUSH A FEW PROPORTIONS THROUGH ITS INNARDS.

BY THE WAY, WHERE'D YOU GET THE HAT?

RESULTS HERE

SUPPOSE TWO RECTANGLES HAVE PROPORTIONAL SIDES a, b, A, B, WITH A = ta AND B = tb (t>0).

a

b

$A = ta$

$B = tb$

THEN

$$\frac{A}{a} = \frac{B}{b} \quad (=t)$$

BUT ALSO

$$\frac{B}{A} = \frac{tb}{ta} = \frac{b}{a}$$

IN OTHER WORDS, ONE PROPORTION IMPLIES ANOTHER. WE ALSO HAVE

$$\frac{B}{A} = \frac{b}{a}$$

THIS IS SOMETHING NEW! A/a WAS THE SCALING FACTOR; WHAT'S b/a?

THE FRACTIONS b/a AND B/A ARE CALLED THE **ASPECT RATIOS** OF THE RECTANGLES, THE RATIO OF **WIDTH** TO **HEIGHT**.

HM... SOUNDS FAMILIAR...

YOU MAY HAVE HEARD THE TERM USED ABOUT TVS, SMARTPHONES, AND COMPUTER DISPLAYS. DESPITE THEIR DIFFERENT SIZES, MANY SCREENS' ASPECT RATIO IS 16:9 ("SIXTEEN TO NINE"). THE WIDTH DIVIDED BY THE HEIGHT IS 16/9 ON ALL OF THEM.

WE HAVE JUST SEEN THAT **PROPORTIONAL RECTANGLES** HAVE **EQUAL ASPECT RATIOS**.

$$\frac{A}{a} = \frac{B}{b}$$

$$\Leftrightarrow$$

$$\frac{b}{a} = \frac{B}{A}$$

SOMETIMES, INSTEAD OF THE SIDES OF A RECTANGLE, a AND b MIGHT BE PARTS OF A LINE SEGMENT. HERE THE POINT Q DIVIDES PR AND WE HAVE THE RATIO

$$\frac{PQ}{QR} = \frac{a}{b}$$

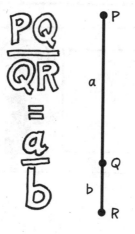

FOR INSTANCE, IF a=6 AND b=2, THE PARTS ARE IN THE RATIO 3:1.

$$\frac{PQ}{QR} = \frac{6}{2} = \frac{3}{1}$$

TWO DIFFERENT SEGMENTS MAY BE DIVIDED PROPORTIONALLY.

MEANING

$$\frac{A}{a} = \frac{B}{b}$$ (THE PARTS SCALE EQUALLY.)

OR

$$\frac{A}{B} = \frac{a}{b}$$ (THE PARTS OF EACH SEGMENT HAVE THE SAME INTERNAL RELATIONSHIP.)

A LITTLE ALGEBRA CRANKS OUT TWO MORE RESULTS WORTH RE-MEMBERING.

ASSUME THAT a, b, A, AND B ARE ALL NONZERO, AND $a/b = A/B$. THEN

$$\frac{b}{a} = \frac{B}{A}$$ (1)

THIS IS PRETTY OBVIOUS. IF x=y, THEN 1/x = 1/y. MORE COMPLICATED IS

$$\frac{a}{a+b} = \frac{A}{A+B}$$ (2)

BECAUSE

$$\frac{a}{b} + 1 = \frac{A}{B} + 1$$

$$\frac{b}{a} + \frac{a}{a} = \frac{A}{B} + \frac{B}{B}$$

$$\frac{a+b}{a} = \frac{A+B}{A}$$

$$\frac{a}{a+b} = \frac{A}{A+B}$$ BY (1)

EQUATION 2 IS NOT REALLY SO MYS-TERIOUS. IF TWO SEGMENTS ARE IN THE RATIO 3:1, SAY, THEN THEY ARE BOTH 3/4 AND 1/4 OF THE TOTAL (3+1 = 4).

$$\frac{9}{3} = \frac{6}{2} = 3 \qquad \frac{9}{12} = \frac{6}{8} = \frac{3}{4}$$

RECALL (P.157) THAT PRO-
PORTIONAL IMAGES HAVE
EQUAL ANGLES EVERY-
WHERE.

WE MIGHT ASK: WHAT IF ONLY
THE CORRESPONDING **VERTEX**
ANGLES ARE EQUAL? WOULD
THAT ENSURE PROPORTIONAL
SIDES?

I'M GUESSING MAYBE!

ANSWER: IN QUADRILATERALS,
NO. TWO RECTANGLES, FOR
INSTANCE, HAVE ALL VERTICES
EQUAL TO 90°, REGARDLESS
OF ASPECT RATIO.

OOPS!

AH, BUT TRIANGLES REALLY ARE DIFFERENT. OVER THE NEXT
FEW PAGES, WE WILL SHOW THAT TRIANGLES **WITH EQUAL
CORRESPONDING VERTEX ANGLES ALSO HAVE
PROPORTIONAL SIDES.**

$$\frac{BP}{BA} = \frac{BQ}{BC}$$

THE ARGUMENT BEGINS WITH THIS
LITTLE "PRE-THEOREM," OR **LEMMA.**

Lemma 15-1. GIVEN △ABC, P BE-
TWEEN A AND B, Q BETWEEN A AND C, IF
PQ∥BC, THEN THE AREAS OF △BPQ AND
△CQP ARE **EQUAL.**

Proof.

1. △BPQ AND △CQP   (DEFS.)
   BOTH HAVE BASE PQ
   AND APEX ON $\overline{BC}$.

2. PQ∥BC   (GIVEN)

3. $\mathcal{A}(\triangle BPQ) = \mathcal{A}(\triangle CQP)$ ▐   (THM. 13-3)

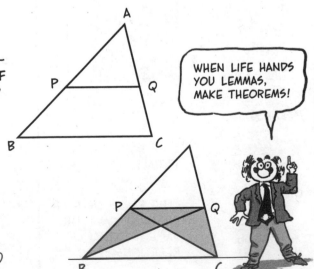

WHEN LIFE HANDS
YOU LEMMAS,
MAKE THEOREMS!

162

Theorem 15-1 (THE SIDE SPLITTER). IF A LINE INTERSECTS TWO SIDES OF A TRIANGLE AT TWO DIFFERENT POINTS AND IS PARALLEL TO THE BASE, THEN IT DIVIDES THE SIDES PROPORTIONALLY.

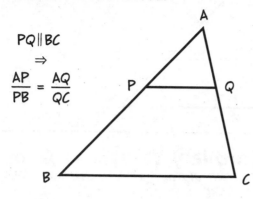

$PQ \parallel BC$

$\Rightarrow$

$$\frac{AP}{PB} = \frac{AQ}{QC}$$

SIDE-SPLITTING? YOUR SIDES? ARE THEY—?

I CAN MUSTER A CHUCKLE...

Proof.

WE SHOW (STEP 3) THAT THE RATIO BETWEEN THE SEGMENTS' PARTS EQUALS THE RATIO BETWEEN TWO **AREAS.** LET h BE THE DISTANCE FROM Q TO THE LINE AB.

1. DRAW SEGMENT BQ. NOTE THAT h IS THE ALTITUDE OF TWO DIFFERENT TRIANGLES, $\triangle APQ$ AND $\triangle BPQ$. THEIR AREAS ARE...

2. $\mathcal{A}(\triangle APQ) = \frac{1}{2}h(AP)$

   $\mathcal{A}(\triangle BPQ) = \frac{1}{2}h(PB)$

3. THEN $\dfrac{\mathcal{A}(\triangle APQ)}{\mathcal{A}(\triangle BPQ)} = \dfrac{AP}{PB}$

4. NOW DRAW SEGMENT PC, AND LET k BE THE DISTANCE FROM P TO AC.

5. AS ABOVE, WE FIND THAT

   $$\frac{\mathcal{A}(\triangle APQ)}{\mathcal{A}(\triangle CPQ)} = \frac{AQ}{QC}$$

6. BUT $\mathcal{A}(\triangle PBQ) = \mathcal{A}(\triangle PQC)$ BY LEMMA 15-1, SO

   $$\frac{AP}{PB} = \frac{\mathcal{A}(\triangle APQ)}{\mathcal{A}(\triangle PQB)} = \frac{\mathcal{A}(\triangle APQ)}{\mathcal{A}(\triangle PQC)} = \frac{AQ}{QC} \ \blacksquare$$

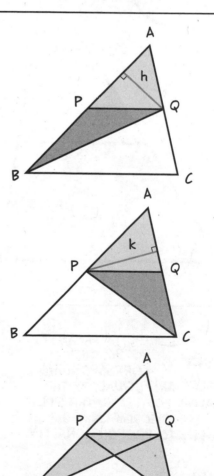

NOW LOOK BACK AT EQUATION (2) ON PAGE 161. IT SAYS THAT IF TWO **PARTS** ARE PROPORTIONAL, THEN THE PARTS ARE ALSO PROPORTIONAL TO THE **WHOLE**.

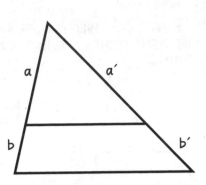

$$\frac{a}{b} = \frac{a'}{b'}$$

$$\Leftrightarrow$$

$$\frac{a}{a+b} = \frac{a'}{a'+b'}$$

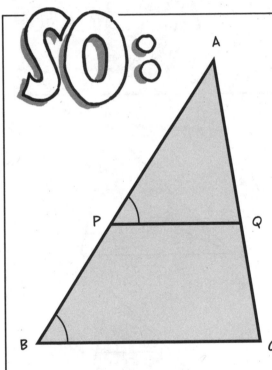

Corollary 15-1.1. IF P AND Q ARE TWO DISTINCT POINTS ON TWO SIDES OF △ABC, AND PQ∥BC, THEN

$$\frac{AP}{AB} = \frac{AQ}{AC}$$

Proof. ASSUME P IS ON AB, Q IS ON AC. SINCE PQ∥BC, THEOREM 15-1 APPLIES.

1. $\dfrac{AP}{PB} = \dfrac{AQ}{QC}$ (THM. 15-1)

2. $\dfrac{AP}{AP+PB} = \dfrac{AQ}{AQ+QC}$ (EQN. 2, P. 161)

3. $\dfrac{AP}{AB} = \dfrac{AQ}{AC}$ ∎   (RULER POST.)

 THE TAKEAWAY: IF PQ∥BC, THEN **CORRESPONDING ANGLES ARE EQUAL**, BY THE PARALLEL POSTULATE. THE SIDE-SPLITTER THEOREM LINKS **EQUAL ANGLES** TO **PROPORTIONALITY**.

$$\angle APQ = \angle B \Rightarrow \frac{AP}{AB} = \frac{AQ}{AC}$$

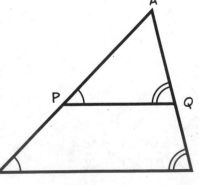

YESSS!

THE COROLLARY JUSTIFIES A NIFTY

CONSTRUCTION,

NAMELY, DIVIDING A LINE SEGMENT INTO **ANY NUMBER OF EQUAL PARTS.** TO ILLUSTRATE, WE DIVIDE A SEGMENT AB INTO THIRDS.

NO RULERS! NO CALCULATORS! NO LONG DIVISION!!

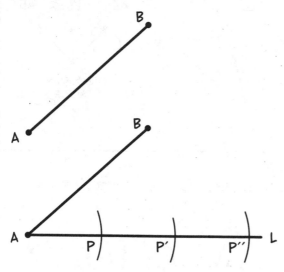

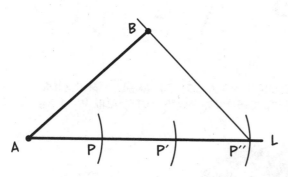

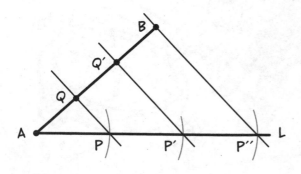

1. DRAW ANY LINE L DIFFERENT FROM $\overline{AB}$ THROUGH POINT A.

2. ON L MARK THREE POINTS P, P', AND P'' AT EQUAL INTERVALS FROM A, THAT IS, SO THAT AP = PP' = P'P''. (THESE CAN BE OF ANY LENGTH!)

3. DRAW P''B.

4. DRAW LINES THROUGH P AND P' PARALLEL TO P''B, INTERSECTING AB AT Q AND Q', RESPECTIVELY. THESE ARE THE POINTS!

WHY? BECAUSE OF PROPORTIONALITY!

BY COROLLARY 15-1.1,

$$\frac{AQ}{AB} = \frac{AP}{AP'} = \frac{1}{3}$$

$$\frac{AQ'}{AB} = \frac{AP'}{AP''} = \frac{2}{3}$$

SO Q AND Q' ARE $\frac{1}{3}$ AND $\frac{2}{3}$ OF THE DISTANCE FROM A TO B.

THERE'S A WORD TO DESCRIBE TWO FIGURES WITH THE SAME SHAPE:

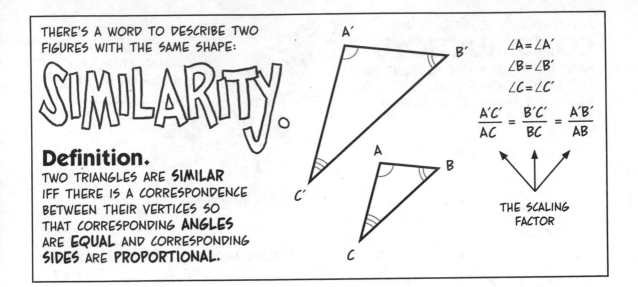

SiMiLARiTY.

Definition.

TWO TRIANGLES ARE **SIMILAR** IFF THERE IS A CORRESPONDENCE BETWEEN THEIR VERTICES SO THAT CORRESPONDING **ANGLES** ARE **EQUAL** AND CORRESPONDING **SIDES** ARE **PROPORTIONAL.**

$$\angle A = \angle A'$$
$$\angle B = \angle B'$$
$$\angle C = \angle C'$$

$$\frac{A'C'}{AC} = \frac{B'C'}{BC} = \frac{A'B'}{AB}$$

THE SCALING FACTOR

THE SYMBOL FOR SIMILARITY $\sim$ IS LIKE THE CONGRUENCE SIGN $\cong$ STRIPPED OF THE PART THAT SUGGESTS EQUALITY. WE WRITE:

$$\triangle ABC \sim \triangle A'B'C'$$

PROVING SIMILARITY WOULD SEEM TO INVOLVE CHECKING A LOT OF ANGLES AND SIDES, BUT (AS WITH CONGRUENCE) THERE ARE SOME TIME-SAVING SHORTCUTS, AND WE LIKE CUTTING TIME.

OTHERWISE, IT WOULD GET AWAY FROM US!

HERE'S AN EXAMPLE OF A TIME-SAVING THEOREM:

Theorem 15-2. IF TWO ANGLES OF ONE TRIANGLE ARE EQUAL TO TWO ANGLES OF ANOTHER, THEN THE TRIANGLES ARE SIMILAR.

Proof. WE ASSUME THAT $\angle A = \angle A'$, $\angle B = \angle B'$, AND THAT $A'B' < AB$, I.E., THE "PRIME" TRIANGLE IS SMALLER. WE THEN "TUCK" A COPY OF $\triangle A'B'C'$ INTO A VERTEX OF $\triangle ABC$.

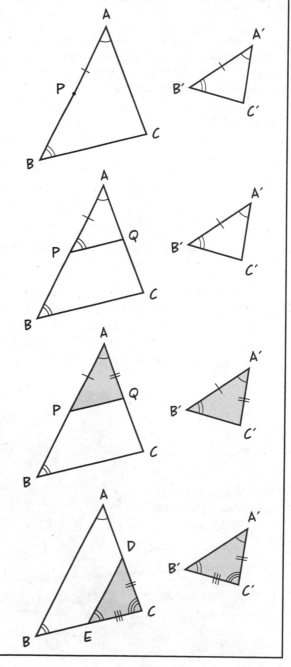

1. ON AB, MARK P SO THAT AP = A'B'.

2. DRAW PQ ∥ BC.

3. $\angle B = \angle APQ$ (PAR. POST.)

4. $\angle B' = \angle APQ$ (SUBST.)

5. $\angle A = \angle A'$ (ASSUMED)

6. $\triangle APQ \cong \triangle A'B'C'$ (ASA)

7. $AQ = A'C'$ (CORR. PARTS)

8. $\dfrac{AP}{AB} = \dfrac{AQ}{AC}$ (COR. 15-1.1)

9. $\dfrac{A'B'}{AB} = \dfrac{A'C'}{AC}$ (SUBST.)

THAT TAKES CARE OF TWO SIDES. BUT ALSO $\angle C = \angle C'$ BECAUSE THE OTHER TWO ANGLES ARE EQUAL.

10. DO THE SAME CONSTRUCTION AT VERTEX C. (TAKE CE = B'C'; DRAW DE ∥ AB, SO $\angle CED = \angle B = \angle B'$, SO $\triangle DEC \cong \triangle A'B'C'$, ETC.) THEN BY THE SAME REASONING,

11. $\dfrac{B'C'}{BC} = \dfrac{A'C'}{AC}$

COMBINING 9 AND 11,

12. $\dfrac{A'B'}{AB} = \dfrac{A'C'}{AC} = \dfrac{B'C'}{BC}$

THE THREE ANGLES ARE EQUAL AND ALL THE SIDES ARE PROPORTIONAL, SO THE TRIANGLES ARE SIMILAR. ∎

A REAL-WORLD APPLICATION?!

Finding a Dinosaur's Height

THE EARTH IS SO FAR FROM THE SUN THAT THE SOLAR RAYS WE RECEIVE ARE "AS GOOD AS" PARALLEL.

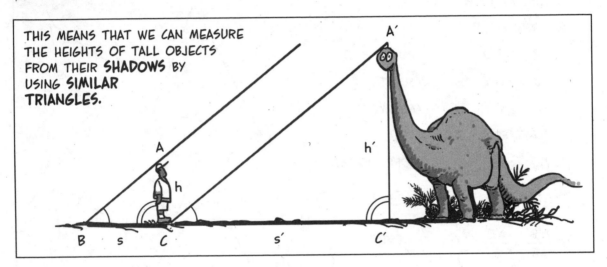

THIS MEANS THAT WE CAN MEASURE THE HEIGHTS OF TALL OBJECTS FROM THEIR **SHADOWS** BY USING **SIMILAR TRIANGLES.**

ASSUMING THE GROUND IS FLAT, $\angle ABC = \angle A'CC'$ BECAUSE THE HYPOTENUSES ARE PARALLEL, AND $\angle ACB = \angle A'C'C$ BECAUSE $AC \parallel A'C'$. TWO ANGLES BEING EQUAL, $\triangle ABC \sim \triangle A'CC'$ BY THEOREM 15-2, SO

$$\frac{h'}{h} = \frac{s'}{s}$$

$$h' = \frac{h}{s} s'$$

AND WE CAN EASILY MEASURE s, s', AND h, OR FAIRLY EASILY.

UH... WHO WANTS TO MEASURE THE DINOSAUR'S SHADOW?

OH... ALL RIGHT...

FOR INSTANCE, WHEN $h = 180\,cm$, $s = 270\,cm$, AND $s' = 930\,cm$,

$$h' = \frac{(180\,cm)(930\,cm)}{270\,cm}$$

$$= 620\,cm$$

$$= \mathbf{6.2\,m}$$

LOOKIT THAT.

SCARED OF HIS OWN SHADOW, AND YET HE TAKES IT WITH HIM...

TWO CONSEQUENCES OF THE TWO-ANGLE TEST:

Corollary 15-2.1. IF TWO TRIANGLES ARE SIMILAR TO A THIRD TRIANGLE, THEN THEY ARE SIMILAR TO EACH OTHER.

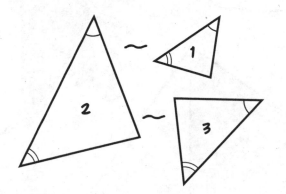

Proof. ASSUME △1∼△2 AND △2∼△3. THEN ALL THE ANGLES OF △1 ARE EQUAL TO ALL THE ANGLES OF △2, WHICH ARE EQUAL TO ALL THE ANGLES OF △3. SO THE ANGLES OF △1 ARE EQUAL TO THE ANGLES OF △3, SO BY THEOREM 15-2, THE TRIANGLES ARE SIMILAR. ∎

AND (MORE SURPRISINGLY)

Theorem 15-3. IN A RIGHT TRIANGLE, THE ALTITUDE FROM THE RIGHT ANGLE TO THE HYPOTENUSE DIVIDES THE TRIANGLE INTO TWO TRIANGLES **SIMILAR TO EACH OTHER AND TO THE ORIGINAL TRIANGLE.**

$\angle ACB = \angle CDB = 90°$
$\Rightarrow$
$\triangle ABC \sim \triangle ACD \sim \triangle CBD$

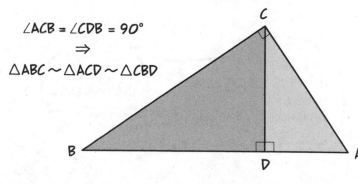

WHERE DOES IT END?

Proof. THIS FOLLOWS STRAIGHT FROM THEOREM 15-2 AND ITS COROLLARY 15-2.1.

1. $\angle A = \angle A$ (OBVIOUS)

2. $\angle ACB = \angle ADC = 90°$ (ASSUMED)

3. $\triangle ACD \sim \triangle ABC$ (TWO ∠s =)

4. LIKEWISE, $\triangle ABC \sim \triangle CBD$

5. THEN $\triangle ACD \sim \triangle CBD$ ∎   (BOTH ∼△ABC)

THESE SIMILARITIES INSIDE A RIGHT TRIANGLE OFFER A FRESH LOOK AT A FAMILIAR THEOREM.

ALSO SOME AMAZING QUILT PATTERNS!

Proof #2 of the Pythagorean Theorem.

LABEL THE SIDES AND VERTICES OF A RIGHT TRIANGLE AS SHOWN.

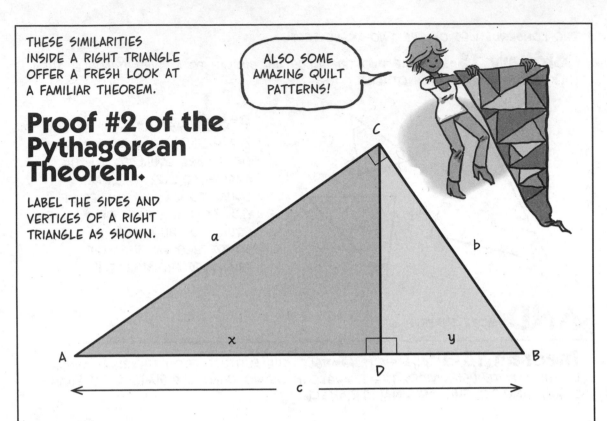

1. BY THEOREM 15-3, $\triangle ABC \sim \triangle ACD \sim \triangle CBD$, SO

$$\frac{y}{b} = \frac{b}{c}, \quad \frac{x}{a} = \frac{a}{c}$$

2. THEN

$$cx = a^2 \quad cy = b^2$$

3. ADDING THESE TWO,

$$cx + cy = a^2 + b^2$$
$$c(x+y) = a^2 + b^2$$

4. BUT $x + y = c$, SO

$$c^2 = a^2 + b^2 \quad \blacksquare$$

I GET IT! a IS THE HYPOTENUSE OF $\triangle CAD$ AND b IS THE HYPOTENUSE OF $\triangle BCD$!!

THEY'RE EVERYWHERE...

TWO TRIANGLES WITH JUST TWO EQUAL ANGLES HAVE PROPORTIONAL SIDES. HOW ABOUT GOING IN THE OTHER DIRECTION? HOW MANY PROPORTIONAL SIDES MUST THERE BE TO IMPLY SIMILARITY?

UM... **IT** DEPENDS!

DO YOU MIND? BEING MORE SPECIFIC, I MEAN?

Theorem 15-4 (SIMILARITY SAS, OR SSAS). GIVEN TWO TRIANGLES, IF TWO PAIRS OF SIDES ARE PROPORTIONAL **AND** THE ANGLES BETWEEN THEM ARE EQUAL, THEN THE TRIANGLES ARE SIMILAR.

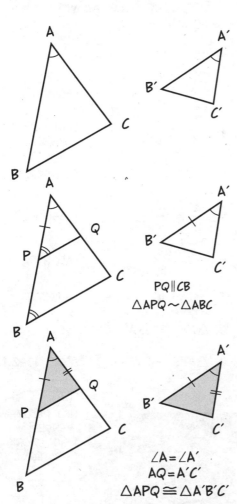

PQ∥CB
△APQ∼△ABC

∠A = ∠A′
AQ = A′C′
△APQ ≅ △A′B′C′

Proof. ASSUMING ∠A = ∠A′ AND

$\dfrac{A'B'}{AB} = \dfrac{A'C'}{AC}$, WE PROVE △ABC∼△A′B′C′.

1. TAKE P ON AB WITH AP = A′B′.   (RULER POST.)

2. DRAW PQ WITH PQ∥BC.   (CONSTR., P. 100)

3. ∠B = ∠APQ   (PAR. POST.)

4. ∠A = ∠A   (OBVIOUS)

5. △APQ∼△ABC   (THM. 15-2)

6. $\dfrac{AQ}{AC} = \dfrac{AP}{AB}$ (BY SIMILARITY)

7. $\dfrac{AQ}{AC} = \dfrac{A'B'}{AB}$ (SUBST.)

8. BUT $\dfrac{A'B'}{AB} = \dfrac{A'C'}{AC}$ (ASSUMED)

9. $\dfrac{AQ}{AC} = \dfrac{A'C'}{AC}$ (SUBST.)

10. AQ = A′C′   (ALGEBRA)

11. △APQ ≅ △A′B′C′   (SAS)

12. △A′B′C′∼△ABC ▮   (BOTH ∼△APQ)

AND ANOTHER TIME-SAVER:

Theorem 15-5 (SIMILARITY SSS OR SSSS). IF TWO TRIANGLES HAVE **ALL THREE SIDES** PROPORTIONAL, THEN THE TRIANGLES ARE SIMILAR.

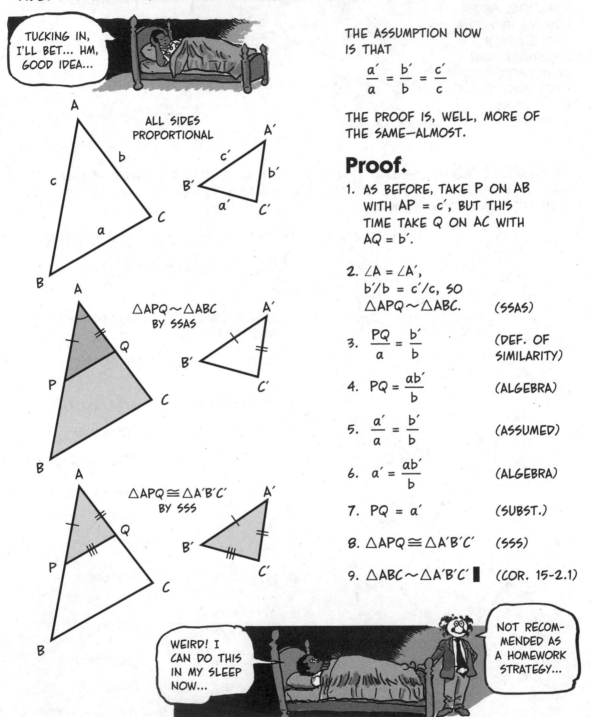

TUCKING IN, I'LL BET... HM, GOOD IDEA...

ALL SIDES PROPORTIONAL

$\triangle APQ \sim \triangle ABC$ BY SSAS

$\triangle APQ \cong \triangle A'B'C'$ BY SSS

THE ASSUMPTION NOW IS THAT

$$\frac{a'}{a} = \frac{b'}{b} = \frac{c'}{c}$$

THE PROOF IS, WELL, MORE OF THE SAME—ALMOST.

Proof.

1. AS BEFORE, TAKE P ON AB WITH AP = c', BUT THIS TIME TAKE Q ON AC WITH AQ = b'.

2. $\angle A = \angle A'$, $b'/b = c'/c$, SO $\triangle APQ \sim \triangle ABC$.   (SSAS)

3. $\dfrac{PQ}{a} = \dfrac{b'}{b}$ (DEF. OF SIMILARITY)

4. $PQ = \dfrac{ab'}{b}$ (ALGEBRA)

5. $\dfrac{a'}{a} = \dfrac{b'}{b}$ (ASSUMED)

6. $a' = \dfrac{ab'}{b}$ (ALGEBRA)

7. $PQ = a'$ (SUBST.)

8. $\triangle APQ \cong \triangle A'B'C'$ (SSS)

9. $\triangle ABC \sim \triangle A'B'C'$ ▮   (COR. 15-2.1)

WEIRD! I CAN DO THIS IN MY SLEEP NOW...

NOT RECOMMENDED AS A HOMEWORK STRATEGY...

172

THESE LAST TWO THEOREMS SHED LIGHT ON THE DISCUSSION AT THE END OF THE LAST CHAPTER.

THERE WE SAW HOW A 3-4-5 RIGHT TRIANGLE SCALES UP TO A 6-8-10 RIGHT TRIANGLE. WE NOW KNOW THAT THESE TWO TRIANGLES ARE **SIMILAR.**

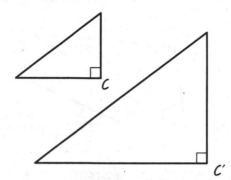

$$\frac{6}{3} = \frac{8}{4} \quad \text{AND} \quad \angle C = \angle C' \quad \text{(SSAS)}$$

$$\frac{6}{3} = \frac{8}{4} = \frac{10}{5} \quad\quad\quad \text{(SSSS)}$$

THIS IMPLIES THAT **ALL** THEIR CORRESPONDING ANGLES MUST BE EQUAL: $\angle A = \angle A'$, $\angle B = \angle B'$. THE SMALL TRIANGLE FITS INTO A CORNER OF THE LARGE ONE, AS WE'VE SEEN.

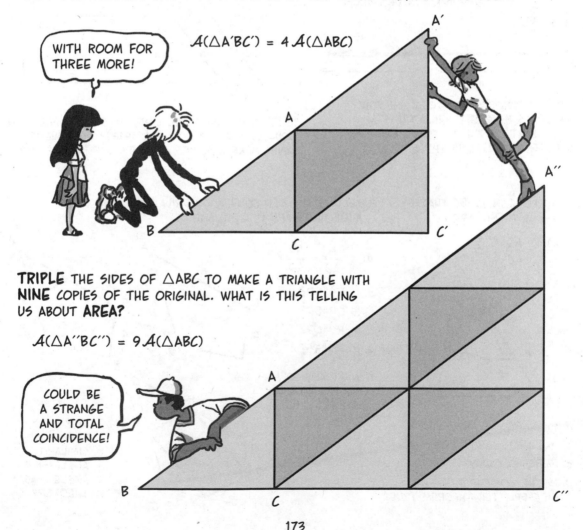

WITH ROOM FOR THREE MORE!

$$\mathcal{A}(\triangle A'BC') = 4\,\mathcal{A}(\triangle ABC)$$

TRIPLE THE SIDES OF $\triangle ABC$ TO MAKE A TRIANGLE WITH **NINE** COPIES OF THE ORIGINAL. WHAT IS THIS TELLING US ABOUT **AREA?**

$$\mathcal{A}(\triangle A''BC'') = 9\,\mathcal{A}(\triangle ABC)$$

COULD BE A STRANGE AND TOTAL COINCIDENCE!

Exercises

1. SOLVE FOR x.

a. $\dfrac{x}{5} = \dfrac{8}{10}$

b. $\dfrac{7}{x} = \dfrac{21}{9}$

c. $\dfrac{3}{x} = \dfrac{x}{27}$

2. ARE THE TWO TRIANGLES SIMILAR?

a.

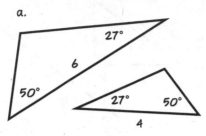

b.

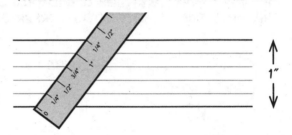

3. SOME GEOMETRY STUDENTS HAVE PUT A STATUE ATOP A 20-FOOT-TALL COLUMN, BUT THEY FORGOT TO MEASURE THE STATUE ITSELF.

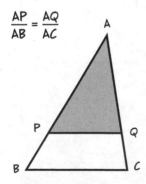

ONE AFTERNOON, THE TOTAL SHADOW MEASURES 39 FEET, WHILE THE STATUE'S PART OF THE SHADOW MEASURES 9 FEET. HOW TALL IS THE STATUE?

4. KEVIN DRAWS PARALLEL LINES ONE INCH APART. HE NOW ROTATES HIS RULER UNTIL IT HAS FIVE QUARTER-INCH TICK MARKS SPANNING THE INCH.

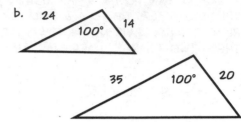

IF HE DRAWS PARALLEL LINES AS SHOWN, ARE THEY EQUALLY SPACED? HOW FAR APART? HOW MIGHT KEVIN MAKE A SERIES OF LINES 3/7 INCH APART?

5. SUPPOSE IN △ABC THAT A-P-B, A-Q-C, AND

$$\dfrac{AP}{AB} = \dfrac{AQ}{AC}$$

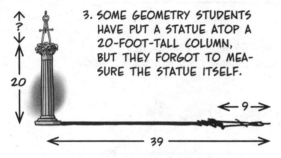

a. IS △APQ ∼ △ABC? WHY?

b. IS ∠APQ = ∠ABC? WHY?

c. IS PQ∥BC? WHY?

d. IS THE CONVERSE OF THE SIDE-SPLITTER THEOREM TRUE?

6. ABCD IS ANY OLD CONVEX QUADRILATERAL. THE SIDES' MIDPOINTS ARE P, Q, R, AND S.

a. IS $\dfrac{AP}{AB} = \dfrac{AS}{AD}$?

b. IS PS∥BD?

c. IS QR∥BD?

d. IS PS∥QR?

e. IS PQ∥RS?

f. WHAT KIND OF QUADRILATERAL IS PQRS?

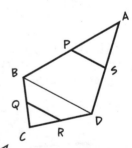

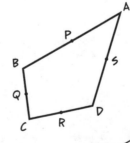

7. WHY IS THERE NO SIMILARITY-ANGLE-SIDE-ANGLE (SASA) THEOREM?

Chapter 16
SCALING AREAS
MORE STUNNING SIMILARITY STUFF

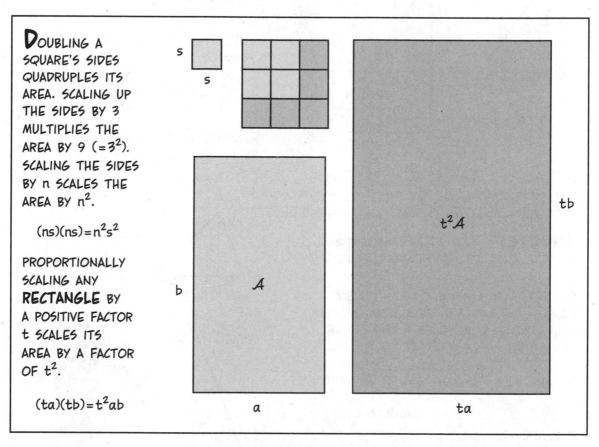

DOUBLING A SQUARE'S SIDES QUADRUPLES ITS AREA. SCALING UP THE SIDES BY 3 MULTIPLIES THE AREA BY 9 ($=3^2$). SCALING THE SIDES BY n SCALES THE AREA BY n^2.

$$(ns)(ns)=n^2s^2$$

PROPORTIONALLY SCALING ANY **RECTANGLE** BY A POSITIVE FACTOR t SCALES ITS AREA BY A FACTOR OF t^2.

$$(ta)(tb)=t^2ab$$

IT WOULD BE STRANGE IF TRIANGLES DIDN'T SCALE THE SAME WAY. AFTER ALL, EVERY TRIANGLE IS HALF A RECTANGLE.

AND TRIANGLES?

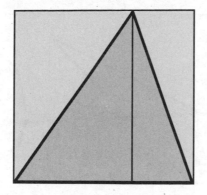

SO YOU'RE SAYING THAT, UM...

EXACTLY!

175

RIGHT TRIANGLES ARE EASY, BECAUSE TWO SIDES ARE ALTITUDES. GIVEN $\triangle ABC \sim \triangle A'B'C'$ WITH $\angle C = \angle C' = 90°$, AND WRITING A FOR $A(\triangle ABC)$ AND A' FOR $A(A'B'C')$, WE HAVE

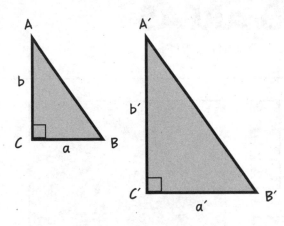

$$A = \tfrac{1}{2}ab \qquad A' = \tfrac{1}{2}a'b' \qquad SO$$

$$\frac{A'}{A} = \frac{a'b'}{ab} = \frac{a'}{a} \cdot \frac{b'}{b}$$

BECAUSE THE TRIANGLES ARE SIMILAR, $b'/b = a'/a$, SO

$$\frac{A'}{A} = \frac{a'}{a} \cdot \frac{a'}{a} = \frac{a'^2}{a^2}$$

THE AREAS ARE PROPORTIONAL TO THE SQUARES OF THE SIDES.

FOR OTHER TRIANGLES, WE FIRST NEED TO SHOW THAT THEIR **ALTITUDES** SCALE LIKE THEIR **SIDES**.

Theorem 16-1. IN SIMILAR TRIANGLES, ALTITUDES ARE PROPORTIONAL TO SIDES.

Proof. WE ASSUME $\triangle ABC \sim \triangle A'B'C'$ AND LET AD AND A'D' BE ALTITUDES FROM CORRESPONDING VERTICES A AND A'. WE SHOW THAT $h'/h = A'C'/AC$.

CORRE-SPONDING ALTITUDES, THAT IS!

1. $\angle ADC = \angle A'D'C' = 90°$ (DEF. OF ALTITUDE)

2. $\angle C = \angle C'$ (SIMILARITY)

3. $\triangle ADC \sim \triangle A'D'C'$ (THM. 15-2)

4. $\dfrac{h'}{h} = \dfrac{A'C'}{AC}$ ∎   (DEF. OF SIM.)

THE ARGUMENT ALSO WORKS IF THE ALTITUDES ARE OUTSIDE THE TRIANGLES.

Theorem 16-2. THE AREAS OF SIMILAR TRIANGLES ARE PROPORTIONAL TO THE SQUARES OF THE SIDES. THAT IS, IF $\triangle ABC \sim \triangle A'B'C'$, THEN $\dfrac{A(\triangle A'B'C')}{A(\triangle ABC)} = \dfrac{a'^2}{a^2}$.

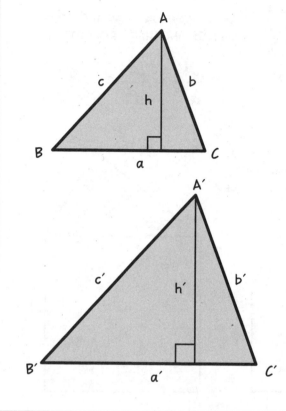

Proof. ASSUME $\triangle ABC \sim \triangle A'B'C'$. LET h AND h′ BE THE RESPECTIVE ALTITUDES, a AND a′ THE RESPECTIVE BASES. THEN

$$\frac{A(\triangle A'B'C')}{A(\triangle ABC)} = \frac{\frac{1}{2}a'h'}{\frac{1}{2}ah} = \frac{a'h'}{ah} = \frac{a'}{a} \cdot \frac{h'}{h}$$

BUT $h'/h = a'/a$, BY THEOREM 16-1, SO

$$\frac{A(\triangle A'B'C')}{A(\triangle ABC)} = \left(\frac{a'}{a}\right)\left(\frac{a'}{a}\right) = \frac{a'^2}{a^2} \quad \blacksquare$$

WRITING t FOR THE SCALING FACTOR a'/a, WE CAN WRITE

$$\frac{A(\triangle A'B'C')}{A(\triangle ABC)} = t^2 \quad \text{OR}$$

$$A(\triangle A'B'C') = t^2 A(\triangle ABC)$$

AGAIN WRITING A FOR $A(\triangle ABC)$ AND A' FOR $A(\triangle A'B'C')$, WE'VE PROVED THE PROPORTION

$$\frac{A'}{A} = \frac{a'^2}{a^2}$$

AS WE SAW ON PAGE 160, THIS IMPLIES ANOTHER PROPORTION:

$$\frac{A'}{a'^2} = \frac{A}{a^2}$$

a^2 IS THE AREA OF THE SQUARE OF SIDE a. THIS PROPORTION SAYS THAT THE **TRIANGLE'S** AREA AND THE **SQUARE'S** AREA **HAVE THE SAME RATIO** WHEN SCALED BY THE SAME FACTOR.

BUT **OF COURSE** THEY DO!

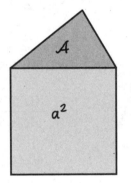

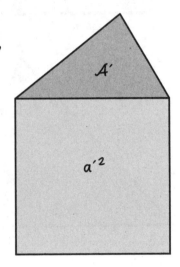

IF THE TRIANGLE HAPPENS TO FIT INSIDE THE SQUARE, WE WOULD SAY THAT THE SIMILAR TRIANGLES TAKE UP THE **SAME FRACTION** OF THEIR CORRESPONDING SQUARES.

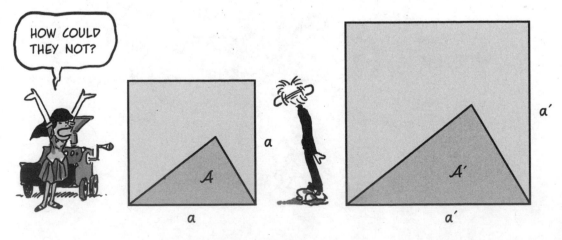

HOW COULD THEY NOT?

Proof #3 of the Pythagorean Theorem.

HERE'S OUR OLD FRIEND, THE RIGHT TRIANGLE △ABC WITH ∠C = 90°, AND AN ALTITUDE CD FROM THE RIGHT ANGLE TO THE HYPOTENUSE. THEN

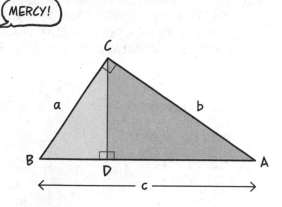

1. $\triangle ADC \sim \triangle CDB \sim \triangle ACB$ (THM. 15-3)

2. THE AREAS ADD:

    $\mathcal{A}(\triangle ADC) + \mathcal{A}\triangle(CDB) = \mathcal{A}(\triangle ACB)$

HERE TWO SIMILAR TRIANGLES' AREAS ADD TO THE AREA OF A THIRD SIMILAR TRIANGLE.

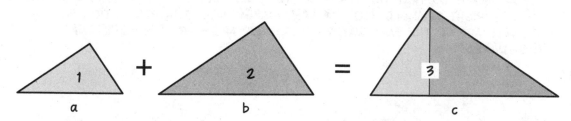

BUT THESE TRIANGLES ALL TAKE UP THE **SAME FRACTION**—CALL IT r—OF THE SQUARES ON THEIR BASES, AS WE JUST SAW. SO—

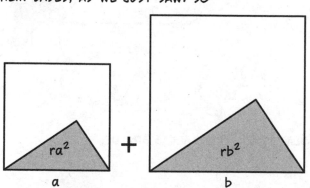

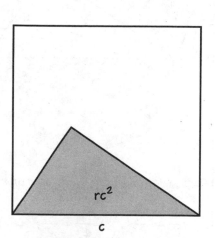

$$\frac{\mathcal{A}(\triangle 1)}{a^2} = \frac{\mathcal{A}(\triangle 2)}{b^2} = \frac{\mathcal{A}(\triangle 3)}{c^2} = \text{SOME NUMBER } r > 0, \text{ SO}$$

$\mathcal{A}(\triangle 1) = ra^2, \ \mathcal{A}(\triangle 2) = rb^2, \ \mathcal{A}(\triangle 3) = rc^2$

$ra^2 + rb^2 = rc^2$ (THE TRIANGLES' AREAS ADD UP.)

$a^2 + b^2 = c^2$ (DIVIDING BY r) ∎

THIS PROOF, MY FAVORITE, SHOWS MOST CLEARLY HOW THE FORMULA ARISES FROM THE SIMILARITIES WITHIN RIGHT TRIANGLES—IN A FLAT PLANE. THE PYTHAGOREAN THEOREM SAYS THAT **THE PLANE IS FLAT.**

IN A CURVED WORLD, THOSE TRIANGLES AREN'T SIMILAR AND THE THEOREM ISN'T TRUE!

IT ALSO REMINDS US THAT THE THEOREM ISN'T ONLY TRUE OF THE **SQUARES** ON THE SIDES OF A RIGHT TRIANGLE. **ANY SIMILAR SHAPES,** SCALED IN PROPORTION TO THE SIDES, WILL ADD UP THE SAME WAY, BECAUSE THEIR AREAS ARE **PROPORTIONAL TO THE SQUARES!**

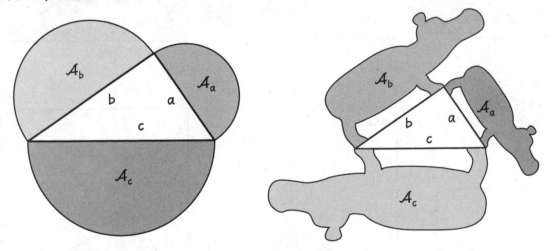

NO MATTER THE SHAPE, IF THEY'RE SIMILAR, THEN THEIR AREAS ADD UP.

$$\frac{\mathcal{A}_b}{\mathcal{A}_a} = \frac{b^2}{a^2}, \qquad \frac{\mathcal{A}_c}{\mathcal{A}_a} = \frac{c^2}{a^2}, \text{ SO}$$

$$a^2 + b^2 = c^2 \qquad \Leftrightarrow$$

$$1 + \frac{b^2}{a^2} = \frac{c^2}{a^2} \qquad \Leftrightarrow$$

$$1 + \frac{\mathcal{A}_b}{\mathcal{A}_a} = \frac{\mathcal{A}_c}{\mathcal{A}_a} \qquad \Leftrightarrow$$

$$\mathcal{A}_a + \mathcal{A}_b = \mathcal{A}_c$$

Example: A Scale Model

MOMO HAS A SCALE MODEL OF HER HOUSE ON A SCALE OF 1:15. THIS MEANS THAT ANY LENGTH IN THE HOUSE, SAY THE HEIGHT OF A WINDOW, IS **15 TIMES** THE CORRESPONDING LENGTH IN THE MODEL.

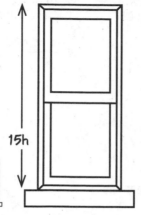

15h

h

MOMO BUYS A SMALL CAN OF HOUSE PAINT AND CAREFULLY PAINTS THE MODEL.

IN THE END, IT TAKES **0.3** LITERS OF PAINT (ABOUT A PINT) TO COVER IT.

HOW MUCH PAINT WILL IT TAKE TO COVER THE WHOLE HOUSE? IT'S A QUESTION OF AREA: THE SURFACE AREA OF THE WALLS. LET A_H = SURFACE AREA OF THE HOUSE, A_M = AREA OF THE MODEL, L_H = SOME LENGTH IN THE HOUSE, L_M = THE CORRESPONDING LENGTH IN THE MODEL.

WE ARE GIVEN

$$\frac{L_H}{L_M} = 15$$

SO

$$\frac{A_H}{A_M} = \left(\frac{L_H}{L_M}\right)^2 = 15^2$$

$$= 225$$

LOVE THE LEOPARD-SKIN HAIR!

PAINT IS PROPORTIONAL TO AREA; THAT IS,

$$\frac{PAINT(HOUSE)}{PAINT(MODEL)} = \frac{A_H}{A_M} = 225$$

$$\frac{PAINT(HOUSE)}{0.3L} = 225$$

$$PAINT(HOUSE) = 225(0.3L) = \mathbf{67.5L}$$

MOMO NEEDS 67.5 LITERS OF PAINT FOR THE WHOLE HOUSE (ABOUT 17 GALLONS).

AND NOW FOR SOMETHING COMPLETELY DIFFERENT!

A LITTLE BIT DIFFERENT, ANYWAY...

BY THE WAY, HOW DID YOU **KNOW** THAT YOU USED 0.3 LITERS ON THE MODEL?

Exercises

1. TWO PHONE SCREENS HAVE THE SAME ASPECT RATIO. ONE MEASURES 5.5" DIAGONALLY; THE OTHER MEASURES 6.2" DIAGONALLY. IF $\mathcal{A}$ IS THE AREA OF THE LARGER AND $\mathcal{A}'$ IS THE AREA OF THE SMALLER, WHAT IS $\mathcal{A}/\mathcal{A}'$? (PRETEND THE SCREENS ARE RECTANGLES WITHOUT CURVED CORNERS.)

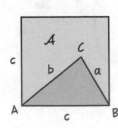

a. ABOUT 127/100

b. ABOUT 1.27

c. 3,844/3,025

d. ALL OF THE ABOVE

e. NONE OF THE ABOVE

f. CAN'T ANSWER WITHOUT KNOWING THE ASPECT RATIO

2. IN △ABC, POINTS P AND Q ARE THE MIDPOINTS OF AB AND AC.

a. $\dfrac{\mathcal{A}(\triangle APQ)}{\mathcal{A}(\triangle ABC)} = ?$ b. $\dfrac{\mathcal{A}(\triangle APQ)}{\mathcal{A}(PQBC)} = ?$

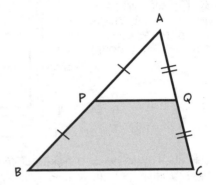

3. SAME QUESTIONS AS IN EXERCISE 2 WHEN

$$\frac{AP}{AB} = \frac{AQ}{AC} = \frac{1}{3}$$

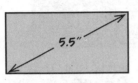

4a. WHAT IS THE RATIO OF THE AREA OF THE LARGE SQUARE TO THE AREA OF THE SMALL SHADED ONE?

4b. WHAT IS THE RATIO BETWEEN THE SMALL SHADED SQUARE AND ONE OF THE SHADED CORNER TRIANGLES?

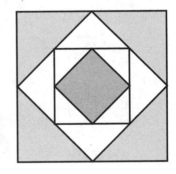

5. SUPPOSE △ABC ∼ △DEF. LET $\mathcal{A}$ = THE AREA OF THE SMALL LIGHT-GRAY SHAPE AND $\mathcal{A}'$ = THE AREA OF THE LARGE LIGHT-GRAY SHAPE. SHOW THAT

$$\frac{\mathcal{A}'}{de} = \frac{\mathcal{A}}{ab}$$

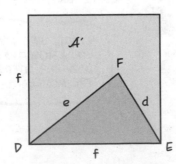

6. AS IN EXERCISE 3, AP/AB = AQ/AC = 1/3. IF R IS THE MIDPOINT OF BC, WHAT IS

$$\frac{\mathcal{A}(\triangle QRC)}{\mathcal{A}(PQRB)} ?$$

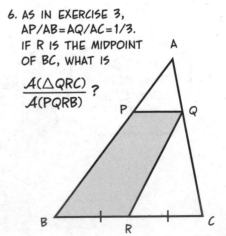

Chapter 17
CIRCLING BACK TO CIRCLES
FINALLY, SOMETHING CURVED, FOR A CHANGE

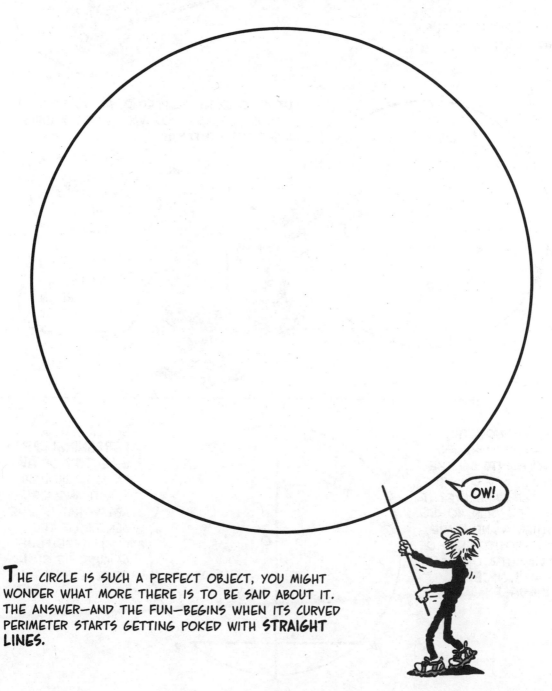

OW!

THE CIRCLE IS SUCH A PERFECT OBJECT, YOU MIGHT WONDER WHAT MORE THERE IS TO BE SAID ABOUT IT. THE ANSWER—AND THE FUN—BEGINS WHEN ITS CURVED PERIMETER STARTS GETTING POKED WITH **STRAIGHT LINES.**

DRAWING ANY CIRCLE STARTS AT THE CENTER O. PLANTING ONE POINT OF THE COMPASS AT O, WE USE THE OTHER TO MARK OFF ALL THE POINTS AT A GIVEN DISTANCE, THE RADIUS, FROM CENTER.

BUT IF SOMEONE SLAPPED DOWN A CIRCLE IN FRONT OF YOU, HOW WOULD YOU KNOW WHERE THE CENTER IS?

EVERY POINT SEEMS ABOUT THE SAME...

LOOK FOR A TACK MARK IN THE PAPER...?

WE DO KNOW THIS: IF A AND B ARE ANY TWO POINTS ON THE CIRCLE, AND O IS THE CENTER (WHEREVER IT MAY BE), THEN AO=BO. EITHER A AND B ARE AT OPPOSITE ENDS OF A DIAMETER, OR △AOB IS AN **ISOSCELES TRIANGLE.**

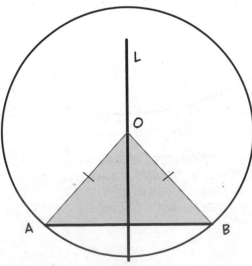

EITHER WAY, A **PERPENDICULAR BISECTOR** OF **AB** PASSES THROUGH THE TRIANGLE'S APEX—THAT IS, THE CENTER OF THE CIRCLE. THE POINT O MUST LIE ON L (COROLLARY 6-3.1).

184

A SEGMENT LIKE AB, WITH TWO ENDPOINTS ON THE CIRCLE, IS CALLED A

CHORD.

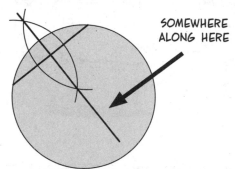

AND WE HAVE JUST PROVED...

Theorem 17-1. A CIRCLE'S CENTER IS ON THE PERPENDICULAR BISECTOR OF ANY CHORD.

SOMEWHERE ALONG HERE

THIS IS TRUE EVEN IF THE CHORD HAPPENS TO BE A DIAMETER, AS THE CENTER IS A DIAMETER'S MIDPOINT.

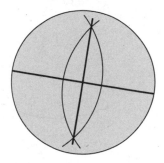

WHERE, EXACTLY, ON THIS LINE IS THE CENTER? HERE ARE TWO WAYS TO FIND IT:

1. TAKE A SECOND CHORD CD, NOT PARALLEL TO AB. THE TWO PERPENDICULAR BISECTORS INTERSECT AT THE CENTER. (COMPARE THIS TO THE CONSTRUCTION ON P. 87.)

OR

2. MAKE A PERPENDICULAR BISECTOR OF AB. EXTEND IT TO A DIAMETER AND FIND ITS MIDPOINT.

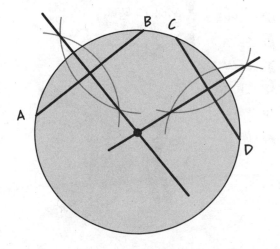

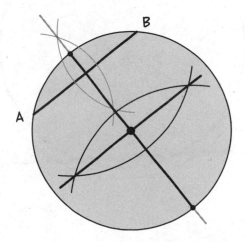

Angles in a Circle

A **CENTRAL ANGLE** IS AN ANGLE WITH ITS VERTEX AT THE CIRCLE'S CENTER. SUPPOSE THE ANGLE'S RAYS INTERSECT THE CIRCLE AT TWO POINTS A AND B.

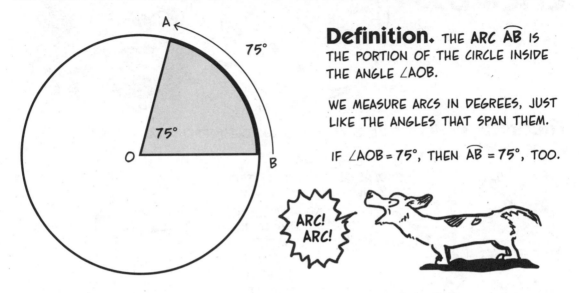

Definition. THE ARC $\overset{\frown}{AB}$ IS THE PORTION OF THE CIRCLE INSIDE THE ANGLE ∠AOB.

WE MEASURE ARCS IN DEGREES, JUST LIKE THE ANGLES THAT SPAN THEM.

IF ∠AOB = 75°, THEN $\overset{\frown}{AB}$ = 75°, TOO.

ARC! ARC!

SINCE A CIRCLE GOES "ALL THE WAY AROUND," WE ALSO INCLUDE ARCS **OUTSIDE** THE ANGLE ∠AOB. THESE ARCS MEASURE **MORE** THAN 180°, LIKE $\overset{\frown}{AB}$ HERE.

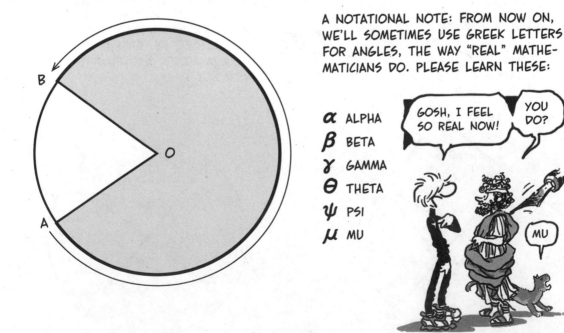

A NOTATIONAL NOTE: FROM NOW ON, WE'LL SOMETIMES USE GREEK LETTERS FOR ANGLES, THE WAY "REAL" MATHEMATICIANS DO. PLEASE LEARN THESE:

α ALPHA
β BETA
γ GAMMA
θ THETA
ψ PSI
μ MU

GOSH, I FEEL SO REAL NOW!

YOU DO?

MU

CHORDS CAN MAKE ANGLES, TOO, IF THEY SHARE AN ENDPOINT. THESE ARE CALLED

INSCRIBED ANGLES.

AN INSCRIBED ANGLE ALSO SPANS AN ARC.

STRANGE WAY TO CUT A PIE.

DEPENDS ON WHAT PIECE YOU WANT...

IF ONE OF THOSE CHORDS PASSES THROUGH THE CENTER, WE CAN DIRECTLY COMPARE THE ARC'S CENTRAL ANGLE WITH ITS INSCRIBED ANGLE.

$\triangle AOC$ IS ISOSCELES, SO $\angle OAC = \angle OCA = \alpha$. THE THIRD ANGLE IS $\theta = 180° - 2\alpha$.

BUT $\angle COA$ AND $\angle AOB$ ARE A LINEAR PAIR, SO ALSO $\theta = 180° - \mu$.

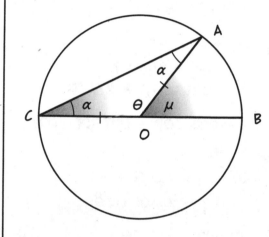

GREEK LETTERS!

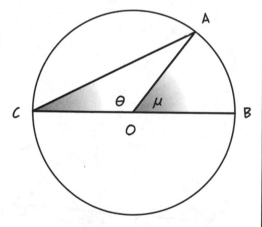

IT FOLLOWS THAT

$$180° - \mu = 180° - 2\alpha$$

SO

$$\mu = 2\alpha$$

THE CENTRAL ANGLE IS TWICE THE INSCRIBED ANGLE.

AT LEAST WHEN ONE CHORD IS A DIAMETER, RIGHT?

WELL, IN FACT...

ANGLES HAVE A WAY OF ADDING UP...
WHICH ENABLES US TO COMPARE CENTRAL
ANGLES WITH INSCRIBED ANGLES EVEN IF
NEITHER CHORD IS A DIAMETER.

Theorem 17-2. ANY INSCRIBED ANGLE MEASURES HALF THE ARC OF THE CORRESPONDING CENTRAL ANGLE.

Proof. GIVEN CHORDS AC AND BC, WITH CENTRAL ANGLE $\angle AOB$, WE SHOW $\angle AOB = 2\angle ACB$. THERE ARE TWO POSSIBILITIES: THE DIAMETER COD IS EITHER INSIDE OR OUTSIDE THE INSCRIBED ANGLE.

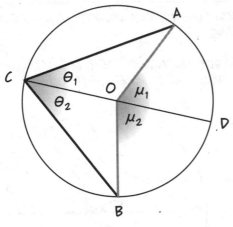

1. FIRST ASSUME THAT DIAMETER COD IS **INSIDE** THE CENTRAL ANGLE $\angle AOB$.

2. LABEL THE ANGLES μ_1, μ_2, θ_1, θ_2, AS SHOWN. AS WE SAW ON PAGE 187,

$$\mu_1 = 2\theta_1 \quad \mu_2 = 2\theta_2$$

3. $\angle AOB = \mu_1 + \mu_2$
   $$= 2\theta_1 + 2\theta_2$$
   $$= 2(\theta_1 + \theta_2) = \mathbf{2\angle ACB}$$

4. IF DIAMETER COD IS **OUTSIDE** $\angle AOB$,

$$\mu_1 = 2\theta_1$$
$$\angle AOB + \mu_1 = 2(\angle ACB + \theta_1)$$
$$= 2\angle ACB + 2\theta_1$$

SUBSTITUTING μ_1 FOR $2\theta_1$,

$$\angle AOB + \mu_1 = 2\angle ACB + \mu_1 \text{ SO}$$

$$\mathbf{\angle AOB = 2\angle ACB} \ \blacksquare$$

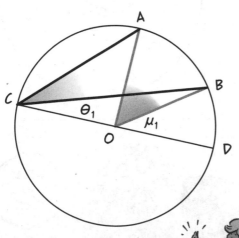

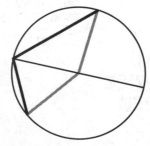

THE THEOREM
IS STILL TRUE
WHEN THE ARC
EXCEEDS 180°.

HERE'S TO
ADDITION AND
SUBTRACTION!

Corollary 17-2.1. ALL INSCRIBED ANGLES SPANNING THE SAME ARC ARE EQUAL.

Proof. THEY'RE ALL EQUAL TO HALF THE CENTRAL ANGLE OF THE ARC. ∎

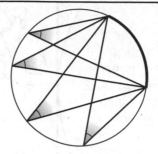

WHO WOULD HAVE THOUGHT?

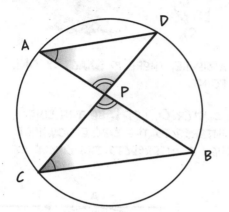

Corollary 17-2.2. IF TWO CHORDS AB AND CD INTERSECT AT A POINT P, THEN △APD ~ △CPB.

Proof. BY COR. 17-2.1, ∠A = ∠C, AND ∠APD = ∠CPB AS VERTICAL ANGLES. SINCE TWO ANGLES ARE EQUAL, THE TRIANGLES ARE SIMILAR. ∎

Corollary 17-2.3. IF TWO CHORDS AB AND CD INTERSECT AT P, THEN

$$AP \cdot PB = CP \cdot PD$$

Proof. BY COR. 17-2.2, △APC ~ △DPB, SO THEIR SIDES ARE PROPORTIONAL.

$$\frac{AP}{PD} = \frac{CP}{PB}$$

CROSS-MULTIPLYING GIVES THE RESULT. ∎

Corollary 17-2.4 (THALES'S THEOREM). IF AB IS A DIAMETER, AND C IS ANY OTHER POINT ON THE CIRCLE, THEN △ABC IS A **RIGHT TRIANGLE.**

Proof. ∠ACB SPANS AN ARC OF 180°, SO ∠ACB = 180°/2 = 90°. ∎

WOW AGAIN!

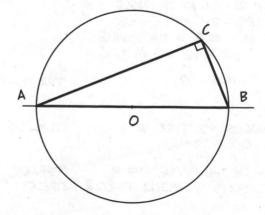

189

Tangents

ARE THERE LINES THAT TOUCH THE CIRCLE AT ONLY ONE POINT? LINES THAT "KISS" IT, AS MATHEMATICIANS SAY? (ACTUALLY, THEY SAY "OSCULATE," BECAUSE IT HAS MORE SYLLABLES.)

Theorem 17-3. THROUGH EACH POINT ON A CIRCLE, THERE IS **EXACTLY ONE LINE** THAT INTERSECTS THE CIRCLE AT ONLY THAT POINT.

Proof. LET A BE A POINT ON THE CIRCLE WITH CENTER O. LET L BE THE LINE THROUGH A WITH L⊥OA. FIRST WE SHOW THAT L INTERSECTS THE CIRCLE NOWHERE ELSE; THEN WE SHOW THAT ANY OTHER LINE THROUGH A INTERSECTS THE CIRCLE AT ANOTHER POINT.

1. LET **B** BE ANY OTHER POINT ON L.   (RULER POST.)

2. L⊥OA   (ASSUMED)

3. △OAB IS A RIGHT TRIANGLE WITH OB AS HYPOTENUSE.   (DEFINITIONS)

4. OB > OA   (THM. 7-3)

5. B IS NOT ON THE CIRCLE.   (DEF. OF CIRCLE)

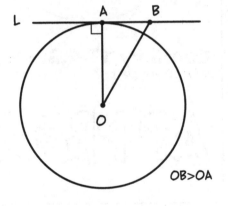

OB>OA

NOW LET M BE A LINE THROUGH A **NOT** PERPENDICULAR TO OA. M AND OA FORM AN ANGLE $\theta < 90°$ ON ONE SIDE.

6. DRAW RAY $\overline{OC}$ AT AN ANGLE $\mu = 180° - 2\theta$ WITH OA ON THE SAME SIDE OF THE RADIUS AS ∠θ, INTERSECTING M AT C.

7. $\angle ACO = 180° - (\theta + \mu)$ (THM. 10-4)
   $\qquad = \theta$

8. △AOC IS ISOSCELES, WITH OA = OC.   (THM. 6-2)

9. C IS ON THE CIRCLE, AND M INTERSECTS THE CIRCLE TWICE. ∎   (DEF. OF CIRCLE)

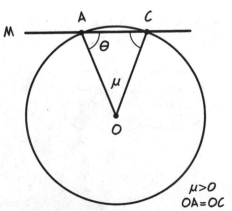

$\mu > 0$
OA = OC

Definition. A LINE THAT TOUCHES A CIRCLE AT A SINGLE POINT IS CALLED A **TANGENT** TO THE CIRCLE.

THEOREM 17-3 SAYS THAT THERE IS EXACTLY ONE "KISSING" TANGENT AT EVERY POINT OF THIS LUCKY CIRCLE: THE PERPENDICULAR TO THE RADIUS AT THAT POINT.

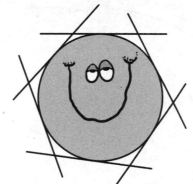

THE NEXT THEOREM SAYS HOW TO "AIM" A TANGENT FROM ANY POINT IN THE PLANE OUTSIDE THE CIRCLE.

Theorem 17-4. GIVEN A POINT P OUTSIDE A CIRCLE C, THERE ARE TWO TANGENTS CONTAINING P.

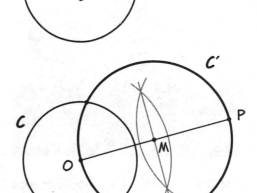

Proof.

1. FIND O, THE CENTER OF C, AND DRAW OP. (CONSTR., P. 87; POST. 2)

2. FIND M, THE MIDPOINT OF OP. (CONSTR., P. 85)

3. DRAW A CIRCLE C' CENTERED AT M WITH RADIUS OM. (DEF. OF CIRCLE)

4. OMP IS A DIAMETER OF C'. (DEF. OF DIAMETER)

5. LET C AND C' INTERSECT AT POINTS Q AND R. (POST. 6)

6. ∠OQP AND ∠ORP ARE RIGHT ANGLES. (COR. 17-2.4)

7. PR AND QR ARE TANGENT TO C. ∎ (THM. 17-3)

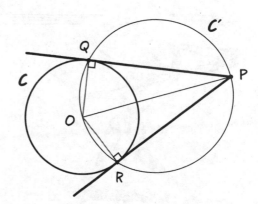

Real-Life Tangents

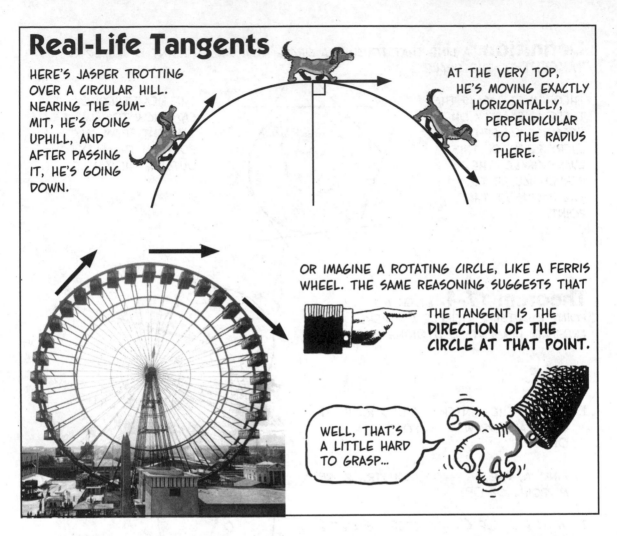

HERE'S JASPER TROTTING OVER A CIRCULAR HILL. NEARING THE SUMMIT, HE'S GOING UPHILL, AND AFTER PASSING IT, HE'S GOING DOWN.

AT THE VERY TOP, HE'S MOVING EXACTLY HORIZONTALLY, PERPENDICULAR TO THE RADIUS THERE.

OR IMAGINE A ROTATING CIRCLE, LIKE A FERRIS WHEEL. THE SAME REASONING SUGGESTS THAT

THE TANGENT IS THE **DIRECTION OF THE CIRCLE AT THAT POINT.**

WELL, THAT'S A LITTLE HARD TO GRASP...

IN PRACTICAL TERMS, IT MEANS THIS: IF BIANCA SWINGS HER PURSE AROUND IN A CIRCLE AND THEN LETS IT GO, IT WILL LITERALLY "FLY OFF ON A TANGENT," FOLLOWING THE CIRCLE'S TANGENT LINE AT THE POINT OF RELEASE.

BY THEOREM 17-3, A CIRCLE'S TANGENTS PASS THROUGH EVERY POINT IN THE PLANE OUTSIDE THE CIRCLE, SO WITH GOOD AIM AND A STRONG ARM, BIANCA SHOULD BE ABLE TO HIT ANY TARGET ANYWHERE!

BIANCA

OKAY... WE KIND OF WENT OFF ON A TANGENT OURSELVES FOR A MINUTE...

HEY! GIVE THAT BACK!

MOVING RIGHT ALONG, THEN...

HM... NOTHING HERE BUT A STRAIGHTEDGE AND COMPASS...

Exercises

1. FIND THE UNKNOWN QUANTITIES:

a.
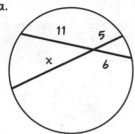

b.
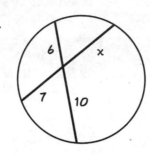

c. HERE AP=PB.
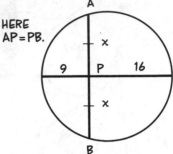

2. THESE TWO CHORDS INTERSECT AT A POINT P OUTSIDE A CIRCLE.

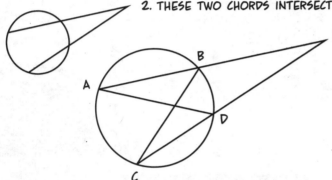

a. DRAW THE CHORDS AD AND BC. IS THERE A PAIR OF SIMILAR TRIANGLES CONTAINING THE VERTEX P? WHY?

b. SHOW THAT PA·PB = PD·PC.

c. WRITING $\overset{\frown}{AC}$ FOR THE ANGLE MEASURE OF THE ARC, SHOW THAT $\angle P = \frac{1}{2}(\overset{\frown}{AC} - \overset{\frown}{BD})$.

3. SUPPOSE P IS ANY POINT ON THE **ANGLE BISECTOR** OF ∠BAC. DRAW PERPENDICULARS PQ AND PR FROM P TO AB AND AC, RESPECTIVELY.

a. WHY IS △AQP ≅ △ARP? CONCLUDE THAT PQ=PR.

b. WHY IS THE CIRCLE CENTERED AT P WITH RADIUS PQ=PR TANGENT TO AB AND AC?

c. IN △ABC, LET THE ANGLE BISECTOR AT B INTERSECT THE ANGLE BISECTOR AT A AT POINT D. WHY IS THE CIRCLE CENTERED AT D, TANGENT TO AB AND AC, ALSO TANGENT TO BC?

d. THE INSCRIBED CIRCLE IS CALLED THE TRIANGLE'S **INCIRCLE**. DOES ITS CONSTRUCTION IMPLY THAT THE ANGLE BISECTORS AT THE THREE VERTICES OF A TRIANGLE INTERSECT AT A POINT?

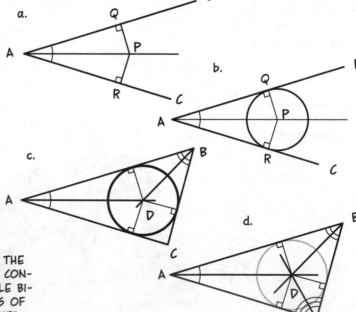

Chapter 18
THE GEOMETRIC MEAN
AN EXCITING NEW WAY TO BE AVERAGE

IN THIS CHAPTER, WE SPEND SOME TIME WITH A DIAMETER BISECTING A CHORD.

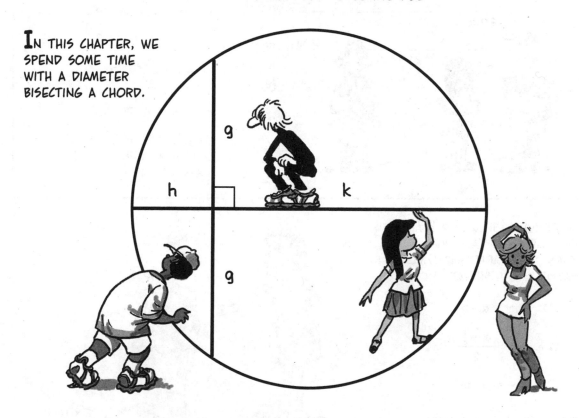

AS YOU CAN PLAINLY SEE, EACH HALF OF THE CHORD, THE LENGTH MARKED g, IS GREATER THAN h AND LESS THAN k, THE PARTS OF THE DIAMETER. WE WILL SHOW THAT g IS A KIND OF "GEOMETRIC AVERAGE" OF h AND k.

AVERAGE... LIKE, ORDINARY, SO-SO, UNREMARKABLE, DULL, NOT EXTREME OR WEIRD?

AHEM... SOMETHING CAN BE REMARKABLE AND NOT BE EXTREME...

THE OLD-FASHIONED **ARITHMETIC** (ACCENT ON "MEH") AVERAGE OF TWO NUMBERS IS THE NUMBER HALFWAY BETWEEN THEM ON THE NUMBER LINE. IF KEVIN IS **180cm** TALL AND MOMO IS **140 cm,** THEN THEIR **MEAN HEIGHT** m IS THE MIDPOINT BETWEEN 180 AND 140. IT'S THE SAME DISTANCE FROM BOTH.

$$180 - m = m - 140$$

$$2m = 180 + 140$$

$$m = \frac{320}{2}$$

$$m = 160 \text{ cm}$$

BY COINCIDENCE, PROFESSOR G HAPPENS TO BE 160 cm TALL.

IN GENERAL, IF a AND b ARE ANY TWO NUMBERS (ASSUME a < b), THEN THEIR ARITHMETIC MEAN m SATISFIES THESE EQUATIONS:

(1) $2m = a + b$

(2) $m = \dfrac{a + b}{2}$

(3) $b - m = m - a$

EQUATION 1 SAYS THAT ADDING THE TWO NUMBERS IS THE SAME AS **ADDING THE MEAN TWICE.**

$$m + m = a + b$$

$$\begin{array}{c} 2 \times 160 \\ = \\ 180 + 140 \end{array}$$

NOW LET'S TRY THE SAME THING WITH **MULTIPLICATION** INSTEAD OF ADDITION. IMAGINE A STRAND OF ELASTIC 1 UNIT LONG TIED AT ONE END TO A STAKE. IMAGINE IT STRETCHED BY A FACTOR OF h, THEN BY A FACTOR OF k. THE RESULT IS **hk** UNITS LONG. FOR INSTANCE, IF h=2 AND k=3, THE ELASTIC STRETCHES TO 6 UNITS LONG.

START   FIRST STRETCH   SECOND STRETCH

WHAT STRETCH FACTOR g WILL PRODUCE THE **SAME LENGTH**, 6, IF DONE **TWICE**?

START   FIRST STRETCH   SECOND STRETCH

IN GENERAL, WE SEE THAT $g^2 = hk$, SO $g = \sqrt{hk}$.

Examples

h=2, k=3 $g = \sqrt{6} \approx 2.449$	h=1, k=2 $g = \sqrt{2} \approx 1.414$	h=0, k=2 $g = \sqrt{0} = 0$	h=5, k=6 $g = \sqrt{30} \approx 5.477$
h=6, k=24 $g = \sqrt{144} = 12$	h=1, k=9 $g = \sqrt{9} = 3$	h=0, k=1,274$\frac{1}{3}$ $g = \sqrt{0} = 0$	h=5, k=125 $g = \sqrt{625} = 25$

THIS NUMBER g, CALLED THE

GEOMETRIC MEAN OF h AND k,

SATISFIES EQUATIONS LIKE THOSE ON THE PREVIOUS PAGE, WITH ADDITION REPLACED BY MULTIPLICATION, SUBTRACTION BY DIVISION, AND HALVING BY TAKING THE SQUARE ROOT.

(4) $g^2 = hk$

(5) $g = \sqrt{hk}$

(6) $\dfrac{k}{g} = \dfrac{g}{h}$

COOL! BUT WHAT'S GEOMETRIC ABOUT IT?

THIS MEAN IS GEOMETRIC BECAUSE IT SOLVES THE PROBLEM OF "SQUARING THE RECTANGLE": FINDING A SQUARE WITH THE SAME AREA AS A GIVEN RECTANGLE.

THIS IS EASY USING ALGEBRA:

$$g^2 = hk$$

$$g = \sqrt{hk}$$

TO FIND g, MULTIPLY h TIMES k AND TAKE THE SQUARE ROOT.

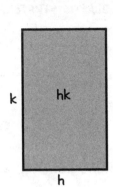

BUT HOW DO I **CON-STRUCT** IT? THAT'S THE QUESTION!

PERHAPS SURPRISINGLY, WE NEED A **CIRCLE** TO DO THIS TRICK.

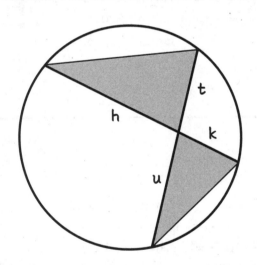

LAST CHAPTER, WE SAW THAT WHEN TWO CHORDS CROSS IN A CIRCLE, THE SHADED TRIANGLES ARE SIMILAR (COR. 17-2.2), GIVING THE PROPORTION

$$\frac{h}{u} = \frac{t}{k}$$

MULTIPLYING BY uk, THIS BECOMES

$$hk = tu \qquad \text{(COR. 17-2.3)}$$

WHAT HAPPENS WHEN **t = u?**

HERE'S A CHORD AB WITH MIDPOINT M. LET g=AM=MB. THEN **ANY OTHER CHORD** PASSING THROUGH M HAS PARTS h AND k THAT SATISFY

$$g^2 = hk$$

CIRCLES. WOW.

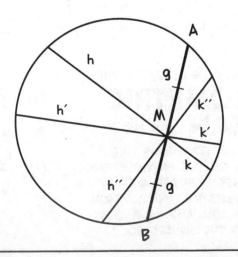

GIVEN A RECTANGLE
WITH SIDES h AND k,
WHAT CIRCLE SHOULD
WE DRAW TO FIND
THEIR GEOMETRIC
MEAN?

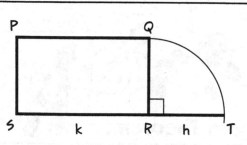

GIVEN RECTANGLE PQRS, EXTEND SR BY
A LENGTH h=QR, MAKING A SEGMENT ST
OF TOTAL LENGTH h + k.

FIND THE MIDPOINT O OF ST, AND DRAW A
CIRCLE CENTERED AT O WITH RADIUS OS=OT.

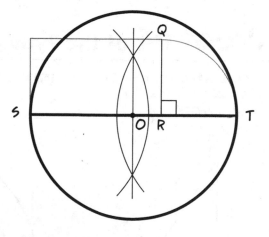

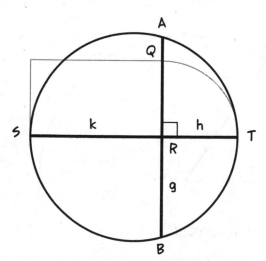

EXTEND SEGMENT RQ, INTERSECTING
THE CIRCLE AT A AND B. THE DIAMETER
ST BISECTS AB (WHY?). LET g = AR = RB.

BY COROLLARY 17-2.3,

$$g^2 = hk$$

EU ROCK!

g IS THE GEOMETRIC
MEAN OF h AND k. THE
SQUARE ON RB HAS THE
SAME AREA AS THE ORI-
GINAL h × k RECTANGLE.

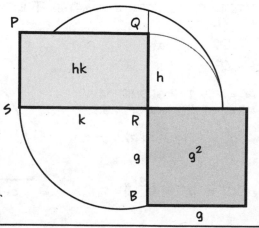

TO SUMMARIZE: TO FIND THE GEOMETRIC MEAN OF h AND k, LAY THE TWO LENGTHS END TO END; DRAW A CIRCLE (A SEMICIRCLE IS ENOUGH) CENTERED AT THE SEGMENT'S MID-POINT AND PASSING THROUGH ITS ENDPOINTS; AND RAISE A PERPENDICULAR AT LENGTH k.

WE'VE ALREADY SEEN SOME RIGHT TRIANGLES IN THAT SEMICIRCLE, AND NOW WE'LL LOOK AT ANOTHER ONE FOR...

OH, YES!

OH, NO!

OH, BOY!

Proof #4 of the Pythagorean Theorem.

HERE'S △ABC WITH SIDES a, b, AND c, ∠C = 90°, c THE HYPOTENUSE.

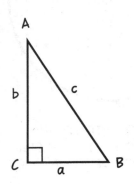

DRAW A CIRCLE CENTERED AT B OF RADIUS c (=AB)

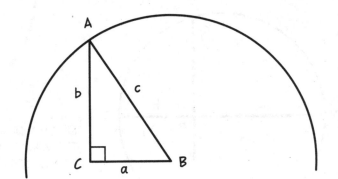

EXTEND CB TO FORM A DIAMETER EF WITH B AT THE MIDPOINT. THEN

$$EC = c - a$$
$$CF = c + a$$

AC = b IS THE GEOMETRIC MEAN OF EC AND CF, I.E.,

$$b^2 = (c-a)(c+a)$$
$$= c^2 - a^2$$
$$c^2 = a^2 + b^2 \ \blacksquare$$

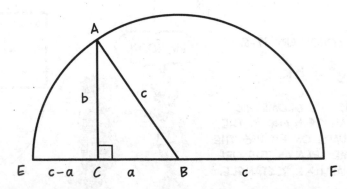

Pythagorean Triples
WARNING: HEAVY ALGEBRA AHEAD!

SKIP THIS SECTION, IF YOU LIKE: NOTHING LATER DEPENDS ON IT.

WHAT? MISS A CHANCE TO TONE MY ALGEBRA MUSCLES?

NO WAY!

LET'S EMPHASIZE THE QUANTITIES WHOSE GEOMETRIC MEAN IS b BY CALLING THEM m AND n.

$$m = c + a \qquad n = c - a$$

THEN

$$m + n = 2c \qquad m - n = 2a \qquad mn = b^2$$

OR

$$c = \frac{m+n}{2} \qquad a = \frac{m-n}{2} \qquad b = \sqrt{mn}$$

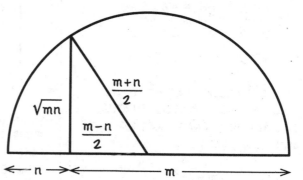

$\sqrt{mn}$ $\frac{m+n}{2}$ $\frac{m-n}{2}$

$\longleftarrow n \longrightarrow$ $\longleftarrow m \longrightarrow$

SUDDENLY WE CAN SEE **WHEN THE TRIANGLE'S SIDES ARE WHOLE NUMBERS.**

IF m AND n ARE BOTH **ODD** (1, 3, 5...), THEN m + n AND m − n ARE BOTH EVEN,* SO c AND a ARE BOTH INTEGERS.

$$c = \frac{m+n}{2}, \quad a = \frac{m-n}{2}$$

IF m AND n ARE BOTH **SQUARES,** SAY $m = p^2$, $n = q^2$, THEN b IS AN INTEGER.

$$b = \sqrt{mn} = \sqrt{p^2 q^2} = pq$$

RESULT: IF m AND n ARE BOTH **ODD SQUARES** (1^2, 3^2, 5^2...) THEN a, b, AND c ARE ALL INTEGERS!

pq

$\frac{p^2 + q^2}{2}$

$\frac{p^2 - q^2}{2}$

Example:

TAKE p = 5, q = 3 (SO m = 25, n = 9)

$$\frac{p^2 + q^2}{2} = \frac{25 + 9}{2} = \frac{34}{2} = \mathbf{17}$$

$$\frac{p^2 - q^2}{2} = \frac{16}{2} = \mathbf{8}$$

$$pq = \mathbf{15}$$

AND WE CHECK:

$$8^2 + 15^2 = 64 + 225$$
$$= 289$$
$$= 17^2$$

*REMEMBER, ODD ± ODD = EVEN.

ANY PAIR OF ODD NUMBERS p, q, WITH $p > q$, GENERATES A PYTHAGOREAN TRIPLE THIS WAY

p	q	pq	$\dfrac{p^2-q^2}{2}$	$\dfrac{p^2+q^2}{2}$
3	1	3	4	5
5	1	5	12	13
7	1	7	24	25
5	3	15	8	17
7	3	21	20	29
9	3	27	36	45
9	5	45	28	53

ETC.

YOU CAN CHECK THAT $a^2 + b^2 = c^2$ LINE BY LINE, OR SLOG THROUGH THE ALGEBRA. AGAIN WRITING $m = p^2$ AND $n = q^2$, WE HAVE

$$a^2 + b^2 = \left(\frac{m-n}{2}\right)^2 + mn$$

$$= \frac{m^2 - 2mn + n^2}{4} + \frac{4mn}{4}$$

$$= \frac{m^2 + 2mn + n^2}{4}$$

$$= \left(\frac{m+n}{2}\right)^2 = c^2$$

TWO LINES ARE HIGHLIGHTED BECAUSE ONE IS A MULTIPLE OF THE OTHER:

$$(27, 36, 45) = 9 \times (3, 4, 5)$$

THAT'S BECAUSE THE GENERATING PAIR $(9,3)$ IS TRIPLE THE PAIR $(3,1)$. 9 AND 3 HAVE A **COMMON FACTOR** OF 3.

THESE TRIANGLES ARE SIMILAR!

YES, THEY ARE.

WHEN p AND q HAVE **NO COMMON FACTOR**, THEIR TRIPLE (a, b, c) ALSO HAS NO COMMON FACTOR AND CAN'T BE "SCALED DOWN" TO A SMALLER PYTHAGOREAN TRIPLE. WE CALL SUCH A TRIPLE **PRIMITIVE.**

NO TWO PRIMITIVE TRIANGLES CAN BE SIMILAR, CAN THEY?

NEVER!

IN FACT, ODD PAIRS p, q WITHOUT COMMON FACTORS GENERATE **ALL** PRIMITIVE PYTHAGOREAN TRIPLES BY THIS FORMULA. **EVERY OTHER PYTHAGOREAN TRIPLE** IS AN INTEGER MULTIPLE OF A PRIMITIVE TRIPLE.*

*YOU CAN FIND A PROOF IN THE WIKIPEDIA ARTICLE ON PYTHAGOREAN TRIPLES.

NOTE: IF YOU'RE INSPIRED OR CONFUSED ENOUGH TO PURSUE THIS FURTHER, YOU'LL RUN INTO **ANOTHER** (SLIGHTLY SIMPLER) FORMULA FOR GENERATING PYTHAGOREAN TRIPLES.

HUH. LIGHTNING STRIKES TWICE?

IT ALSO BEGINS WITH TWO INTEGERS, r AND s, WITH NO COMMON FACTOR, BUT NOW ONE NUMBER MUST BE ODD AND THE OTHER EVEN. ASSUMING $r > s$, THE TRIPLE IS:

$$a = r^2 - s^2$$
$$b = 2rs$$
$$c = r^2 + s^2$$

DOUBLE HUH...

FOR EXAMPLE, $r = 5$, $s = 2$ GIVES

$$a = 25 - 4 = 21$$
$$b = (2)(5)(2) = 20$$
$$c = 25 + 4 = 29$$

ALTHOUGH THIS IS AN EASIER CALCULATION,* DERIVING IT IS JUST A TINY BIT TRICKIER. IF YOU FEEL LIKE TACKLING IT ON YOUR OWN, START WITH THIS DIAGRAM:

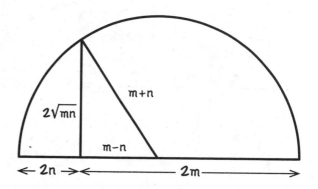

AND LET $m = r^2$, $n = s^2$

HERE ENDS OUR NUMERICAL DETOUR. WE NOW REJOIN THE MAIN ROAD TO MORE MAINSTREAM GEOMETRY.

DID I BREAK THE ALGEBRA ENGINE?

NO... IT JUST NEEDS A TUNE-UP FOR THE NEXT CHAPTER...

*THE ODD-EVEN PAIR (r, s) GIVES THE SAME TRIPLE AS THE ODD-ODD PAIR (r+s, r−s).

Exercises

1. FIND THE SIDE OF EACH SQUARE, ASSUMING THAT ITS AREA IS EQUAL TO THE AREA OF THE RECTANGLE TO ITS LEFT.

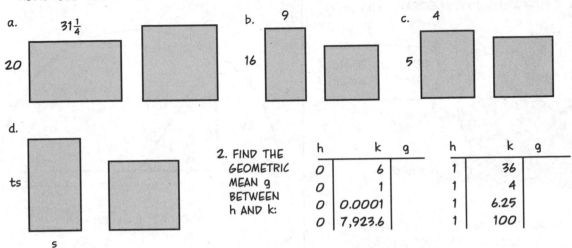

a. $31\frac{1}{4}$

20

b. 9

16

c. 4

5

d.

ts

s

2. FIND THE GEOMETRIC MEAN g BETWEEN h AND k:

h	k	g		h	k	g
0	6			1	36	
0	1			1	4	
0	0.0001			1	6.25	
0	7,923.6			1	100	

3. PROBLEM 2 SUGGESTS A WAY TO FIND SQUARE ROOTS GEOMETRICALLY.

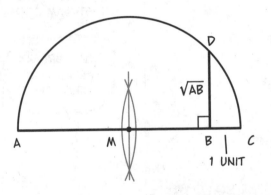

GIVEN ANY NUMBER n>0, DRAW A SEGMENT AB OF LENGTH n, THEN EXTEND IT BY 1 UNIT TO C. DRAW A SEMICIRCLE OF DIAMETER AC, AND RAISE THE PERPENDICULAR BD TO THE CIRCLE. THEN

$$BD = \sqrt{AB}$$

D

$\sqrt{AB}$

A   M   B | C

1 UNIT

TRY THIS CONSTRUCTION WITH AB=4 AND BC=1. MEASURE TO SEE IF DB=2. TRY SOME OTHER VALUES OF n, TOO. WORK LARGE!

4a. IF g IS THE GEOMETRIC MEAN OF h AND k, WHAT IS THE GEOMETRIC MEAN OF 5h AND 5k IN TERMS OF g?

4b. WHAT IS THE GEOMETRIC MEAN OF rh AND rk IN TERMS OF g AND r?

4c. DRAWING THE CONSTRUCTION IN PROBLEM 3 WITH AB=8 AND BC=2 MAKES DB=?

5. THE CONVERSE OF THALES'S THEOREM (P. 189) IS TRUE: WHEN A RIGHT TRIANGLE IS INSCRIBED IN A CIRCLE, THE HYPOTENUSE IS A DIAMETER. (OTHERWISE, THE ARC SPANNED BY ∠C WOULDN'T BE 180°.) DOES THIS IMPLY THAT THE MIDPOINT OF THE HYPOTENUSE IS EQUIDISTANT FROM THE VERTICES? IS IT TRUE THAT MA = MC = MB?

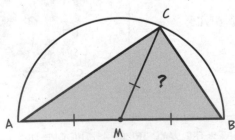

C

?

A   M   B

6. HOW WOULD YOU SQUARE A TRIANGLE, THAT IS, FIND A SQUARE OF EQUAL AREA?

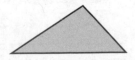

204

Chapter 19
THE GOLDEN TRIANGLE

THE MOST BEAUTIFUL NUMBER, ACCORDING TO SOME PEOPLE

THIS ISOSCELES TRIANGLE, WITH BASE ANGLES 72° AND APEX ANGLE 36°, CONTAINS HIDDEN TREASURES, INCLUDING A HALF-HIDDEN CONNECTION TO THE NUMBER **5**.

$$36° + (2)(36°) + (2)(36°)$$
$$= (5)(36°) = 180°$$

1 × 36°

WONDERFUL NUMBER!

REMINDS ME OF QUITTING TIME...

2 × 36°   2 × 36°

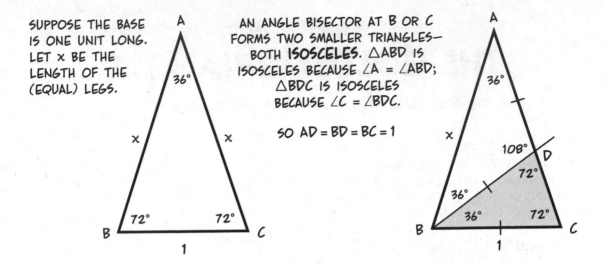

SUPPOSE THE BASE IS ONE UNIT LONG. LET x BE THE LENGTH OF THE (EQUAL) LEGS.

AN ANGLE BISECTOR AT B OR C FORMS TWO SMALLER TRIANGLES—BOTH **ISOSCELES**. △ABD IS ISOSCELES BECAUSE ∠A = ∠ABD; △BDC IS ISOSCELES BECAUSE ∠C = ∠BDC.

SO AD = BD = BC = 1

SINCE △BDC HAS THE SAME ANGLES AS △ABC,

$$\triangle ABC \sim \triangle BDC$$

THIS IMPLIES

$$\frac{AC}{BC} = \frac{BC}{DC}$$

BUT BC=AD, SO

(1) $$\frac{AC}{AD} = \frac{AD}{DC}$$

THE LONG PART AD IS THE **GEOMETRIC MEAN** BETWEEN THE WHOLE SEGMENT AC AND THE SHORT PART DC.

STRANGE?

MAYBE...

NOW AC = x, AD = BC = 1, DC = x−1. IN THOSE TERMS, EQUATION 1 LOOKS LIKE

$$\frac{x}{1} = \frac{1}{x-1} \qquad OR \qquad x(x-1)=1 \quad OR$$

$$x^2 = 1 + x \qquad OR \qquad x^2 - x - 1 = 0$$

WHICH IS QUICKLY SOLVED BY THE **QUADRATIC FORMULA.*** IF YOU DON'T KNOW IT, HERE'S THE SOLUTION:

$$\frac{1+\sqrt{5}}{2}$$

WAWK! THERE'S THAT **FIVE** AGAIN!

STRANGE, AND STRANGE AGAIN...

*IF $x^2 + bx + c = 0$, THEN $x = \frac{1}{2}(-b \pm \sqrt{b^2 - 4c})$. IN GEOMETRY, WE USE ONLY THE POSITIVE ROOT.

206

LET'S CHECK THAT THIS VALUE OF x SATISFIES THE EQUATION.

$$x^2 = \left(\frac{1+\sqrt{5}}{2}\right)^2 = \frac{(1+\sqrt{5})^2}{2^2}$$

$$= \frac{1+2\sqrt{5}+5}{4} = \frac{6+2\sqrt{5}}{4}$$

$$= \frac{4}{4} + \frac{2+2\sqrt{5}}{4} = 1 + \frac{1+\sqrt{5}}{2}$$

$$= 1 + x$$

THIS NUMBER IS SO IMPORTANT THAT IT GETS ITS OWN SYMBOL, THE GREEK LETTER PHI ("FEE" OR "FIE"):

A CALCULATOR THAT TAKES ROOTS GIVES A DECIMAL EXPANSION FOR ϕ:

$$\phi = 1.61803398874985...$$

IT SATISFIES AN EQUATION THAT CAN BE WRITTEN IN VARIOUS WAYS:

(2) $\phi^2 - \phi - 1 = 0$

(3) $\phi^2 = \phi + 1$

(4) $\phi(\phi - 1) = 1$ FACTORING ϕ OUT OF $\phi^2 - \phi$

(5) $\frac{1}{\phi} = \phi - 1$ DIVIDING (4) BY ϕ

(6) $\phi = 1 + \frac{1}{\phi}$ REARRANGING

(7) $1 = \frac{1}{\phi} + \frac{1}{\phi^2}$ DIVIDING (6) BY ϕ

LET'S TRY SOME OF THESE OUT ON THE CALCULATOR.

$$\phi^2 = 2.61803398874985... = \phi + 1$$

$$\frac{1}{\phi} = 0.61803398874985... = \phi - 1$$

$$\frac{1}{\phi^2} = 0.38196601125005... = 1 - \frac{1}{\phi}$$

THIS NUMBER IS **WEIRD!**

OOEE!

ϕ IS CALLED THE **GOLDEN RATIO** BECAUSE IT COMES FROM THE HARMONIOUS RELATIONSHIP BETWEEN THE WHOLE AND THE PARTS EXPRESSED BY EQUATION 1. IT TRULY IS A HIDDEN TREASURE.

AWK! PIECES OF FIVE! PIECES OF FIVE! AWK!

WE HAVE ALL BUT PROVED

Theorem 19-1. IN ANY 36°-72°-72° TRIANGLE, EACH LEG IS φ TIMES THE BASE.
CONVERSELY, IF AN ISOSCELES TRIANGLE HAS A LEG EQUAL TO φ TIMES THE BASE, IT IS A 36°-72°-72° TRIANGLE.

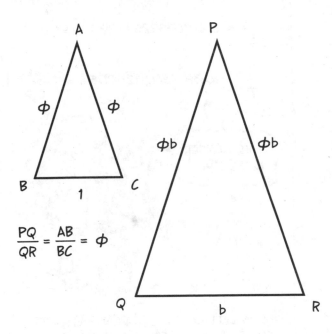

$$\frac{PQ}{QR} = \frac{AB}{BC} = \phi$$

Proof. WE JUST SAW THAT A TRIANGLE △ABC WITH BASE 1 AND BASE ANGLES OF 72° HAS LEGS OF LENGTH φ.

ANY OTHER 36°-72°-72° TRIANGLE △PQR IS SIMILAR TO △ABC BY THEOREM 15-2, SO PQ/QR = AB/BC = φ.

CONVERSELY, IF

$$\frac{PQ}{QR} = \frac{PR}{QR} = \phi$$

THEN △PQR ∼ △ABC BY SSSS, SO CORRESPONDING ANGLES ARE EQUAL. ∎

(INCIDENTALLY, HOW DO WE KNOW THAT A 36°-72°-72° TRIANGLE ACTUALLY EXISTS? BECAUSE: A SEGMENT OF LENGTH 1 EXISTS BY THE RULER POSTULATE. THE 72° CORNER ANGLES EXIST BY THE PROTRACTOR POSTULATE. THE TWO RAYS INTERSECT BY THE PARALLEL POSTULATE, AND MEET AT 36° BY THEOREM 10-4.)

BUT CAN WE **CONSTRUCT** IT? THAT'S THE QUESTION!

OF COURSE WE CAN...

OH, GOOD! YOU SCARED US THERE FOR A MINUTE...

THE CONSTRUCTION
HINGES ON THE
RATIO ϕ, AND ALL
CONSTRUCTIONS
INVOLVING ϕ START
WITH $\sqrt{5}$.

$$\sqrt{1^2+2^2} = \sqrt{5}$$

1

2

$\sqrt{5}$ IS THE HYPOTENUSE OF
A RIGHT TRIANGLE WITH SIDES
1 AND 2.

Theorem 19-2. GIVEN A SEGMENT AB, WE CAN DIVIDE THE SEGMENT AT A POINT E SO THAT AB/EB = ϕ.

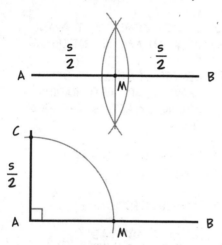

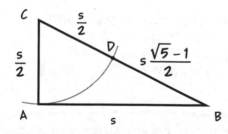

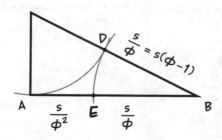

Proof. LET AB = s.

1. FIND THE MIDPOINT M OF AB.

2. DRAW A PERPENDICULAR TO AB
   AT A, AND USING A AS CENTER,
   MARK C ON THE PERPENDICULAR
   WITH AC = AM = s/2.

3. DRAW BC. BY PYTHAGORAS,

$$BC = \sqrt{s^2 + \left(\frac{s}{2}\right)^2} = \sqrt{\frac{5s^2}{4}} = s\frac{\sqrt{5}}{2}$$

$$= s\left(\phi - \frac{1}{2}\right)$$

4. USING C AS CENTER, MARK
   D ON BC WITH CD = CA = s/2.

5. THEN

$$DB = s\left(\frac{\sqrt{5}-1}{2}\right) = s(\phi - 1)$$

6. BUT $\phi - 1 = \dfrac{1}{\phi}$ (EQN. 5), SO

$$DB = \frac{s}{\phi}$$

7. WITH COMPASS CENTERED AT B,
   MARK E ON AB WITH EB = DB. THEN

$$\frac{AB}{EB} = \frac{s}{\left(\frac{s}{\phi}\right)} = \phi \ \blacksquare$$

TO CONSTRUCT A 36°-72°-72° TRIANGLE, WE MAKE AN
ISOSCELES TRIANGLE WHOSE BASE IS 1/ɸ TIMES THE LEG.

Theorem 19-3. GIVEN A SEGMENT AB, A 36°-72°-72° TRIANGLE CAN BE CONSTRUCTED WITH AB AS A LEG.

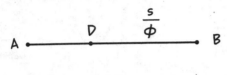

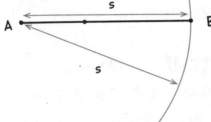

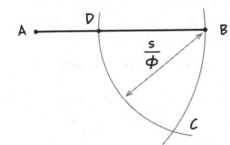

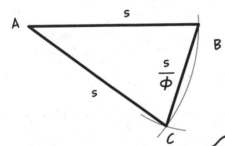

Proof. LET AB = s.

1. BY THEOREM 19-2, DIVIDE AB AT D SO THAT

$$\frac{AB}{DB} = \phi$$

2. CENTER COMPASS AT A AND DRAW AN ARC OF RADIUS AB.

3. CENTER COMPASS AT B AND DRAW AN ARC WITH RADIUS BD, INTERSECTING THE FIRST ARC AT C.

4. BC = BD AND AB = AC BY CONSTRUCTION.

5. BY SUBSTITUTION,

$$\frac{AC}{BC} = \frac{AB}{BC} = \frac{AB}{BD} = \phi$$

6. △ABC IS ISOSCELES, WITH LEG EQUAL TO ɸ × BASE, SO ITS ANGLES ARE 36°, 72°, AND 72° (THM. 19-1). ∎

GREAT! BUT... WHAT'S SO GOLDEN ABOUT IT?

YEAH! CAN I USE IT TO BUY AN EXTRA-LARGE VEGAN PIZZA (PLUS DOUBLE ANCHOVIES)?

IS IT LIKE CRYPTO OR WHAT??

THE ANCIENT GREEKS BELIEVED THAT ϕ WAS THE **ESSENCE OF BEAUTY,** BUT I ASK YOU, A 36°-72°-72° TRIANGLE? SURE, IT'S NICE-ENOUGH LOOKING, BUT I MEAN... REALLY?

WELL, IT ISN'T EXACTLY **UGLY...**

IN FACT, THESE MATHEMATICAL TASTEMAKERS WERE THINKING OF **RECTANGLES,** NOT TRIANGLES. THE **GOLDEN RECTANGLE,** WITH ONE SIDE ϕ TIMES THE OTHER, APPEARS THROUGHOUT WESTERN ART AND ARCHITECTURE—AND ALSO IN NATURAL SPIRAL FORMS LIKE SEASHELLS AND SEED CLUSTERS.

YOU'LL HAVE A CHANCE TO PLAY WITH THIS BOX IN THE EXERCISES...

WHILE I DUST OFF A FEW RIGHT TRI- ANGLES...

Exercises

1. THE GOLDEN RECTANGLE OFFERS A GREAT WAY TO VIEW THE MAGIC OF ϕ. START WITH A $1 \times \phi$ RECTANGLE.

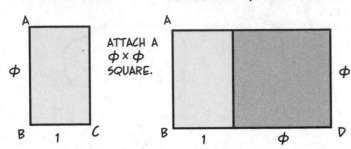

ATTACH A $\phi \times \phi$ SQUARE.

a. SHOW THAT THIS RECTANGLE IS ALSO GOLDEN, THAT IS,

$$\frac{BD}{AB} = \phi$$

(HINT: $\phi + 1 = ?$)

b. ANOTHER ADDED SQUARE MAKES ANOTHER GOLDEN RECTANGLE! FOR:

$$\phi + \phi^2$$
$$= \phi(1 + \phi)$$
$$= \phi \phi^2 = \phi^3$$

SO

$$\frac{AE}{EF} = ?$$

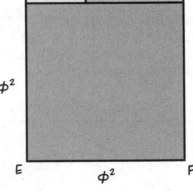

c. SHOW THAT **REMOVING** A 1×1 SQUARE FROM A $1 \times \phi$ RECTANGLE LEAVES ANOTHER GOLDEN RECTANGLE.

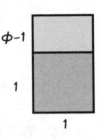

d. SHOW THAT $\phi^{14} + \phi^{15} = \phi^{16}$.

2. HOW LONG IS SEGMENT CE? HOW LONG IS ED?

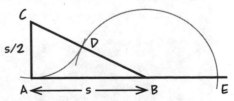

3. IN THIS FIGURE, THE LARGE TRIANGLES LIKE $\triangle ADC$, $\triangle BED$, ETC. HAVE ANGLES OF $36°, 72°, 72°$.

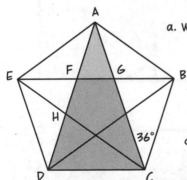

a. WHY IS $\triangle AFG \sim \triangle ADC$?

b. WHAT IS THE RATIO AC/AG?

c. WHY IS $\triangle AEH \sim \triangle ADC$?

d. WHAT IS $\dfrac{\mathcal{A}(\triangle ADC)}{\mathcal{A}(\triangle AFG)}$?

4. IF $AB = s$, HOW LONG IS AE?

5 (SEMI-ADVANCED). THE **FIBONACCI NUMBERS** START WITH 1,1, AND MAKE A NEW TERM BY ADDING THE PREVIOUS TWO: 1, 1, 2, 3, 5, 8, 13, 21, 34, 55, 89, 144, 233, 377, 610, 987... IF F_n IS THE nth FIBONACCI NUMBER, SHOW THAT $\phi^n = F_n \phi + F_{n-1}$. (HINT: KEEP PILING MORE SQUARES ON EVER-LARGER GOLDEN RECTANGLES.)

Chapter 20

TRIG TRICKS

RIGHT TRIANGLES RULE!

In the beginning were lengths and angles, with no apparent connection between them.

AN ANGLE'S SIDE CAN HAVE ANY LENGTH...

AND LENGTHS CAN MEET AT ANY ANGLE!

FOLD BACK ONE OF THOSE RAYS TO MAKE A TRIANGLE, AND SUDDENLY A LINK BETWEEN SIDES AND ANGLES APPEARS: TWO TRIANGLES WITH THE **SAME ANGLES** HAVE **PROPORTIONAL SIDES.** THE RATIO OF ONE SIDE TO ANOTHER IS THE SAME IN ALL LIKE-ANGLED TRIANGLES.

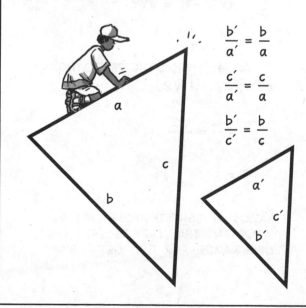

$$\frac{b'}{a'} = \frac{b}{a}$$

$$\frac{c'}{a'} = \frac{c}{a}$$

$$\frac{b'}{c'} = \frac{b}{c}$$

IN THIS CHAPTER, WE LEARN WHAT THOSE RATIOS ARE AND HOW THEY'RE USED TO FIND A TRIANGLE'S **EXACT MEASUREMENTS.**

WE'LL HAVE TAILORED TRIANGLES!

AS USUAL, **RIGHT** TRIANGLES ARE EASIER. IN A RIGHT TRIANGLE, A SINGLE ACUTE ANGLE DETERMINES EVERYTHING. GIVEN ∠A, THEN ∠B = 90° – ∠A. (∠C IS ASSUMED TO BE 90°.)

THAT IS, NO MATTER THE LENGTH OF THE LEGS, IF **ONE** ACUTE ANGLE IS FIXED, THEN THE **RATIOS OF THE SIDES** ARE ALWAYS THE SAME.

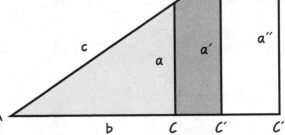

$$\frac{a}{c} = \frac{B'C'}{AB'} = \frac{B''C''}{AB''} = \ldots \text{ETC.}$$

$$\frac{b}{c} = \frac{AC'}{AB'} = \frac{AC''}{AB''} = \ldots \text{ETC.}$$

NEVER ARGUE WITH A TRIANGLE THAT HAS A 90° ANGLE.

WHY NOT?

FOR EXAMPLE, IF ∠A = 30°, WE'VE SEEN THAT $a/c = \frac{1}{2}$, AND THE SIDES ARE IN THE PROPORTION $1 : \sqrt{3} : 2$.

$$\frac{a}{c} = \frac{1}{2}, \quad \frac{b}{c} = \frac{\sqrt{3}}{2}, \quad \frac{b}{a} = \sqrt{3}$$

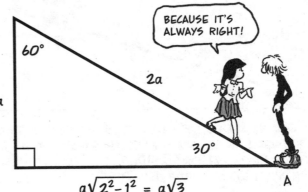

BECAUSE IT'S ALWAYS RIGHT!

$$a\sqrt{2^2 - 1^2} = a\sqrt{3}$$

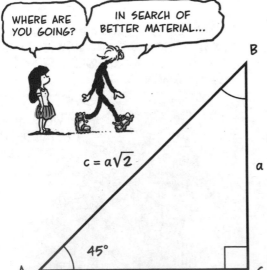

WHERE ARE YOU GOING?

IN SEARCH OF BETTER MATERIAL...

$c = a\sqrt{2}$

IF ∠A = 45°, THEN THE PROPORTIONS $a:b:c$ ARE $1:1:\sqrt{2}$.

$$\frac{a}{c} = \frac{b}{c} = \frac{\sqrt{2}}{2}$$

$$\frac{a}{b} = 1$$

RATIOS IN RIGHT TRIANGLES ARE SO IMPORTANT THAT THEY DESERVE THEIR OWN NAMES—AND THEY GET THEM.

Sine and Cosine

START WITH △ABC, ∠C=90°, LEGS a AND b, HYPOTENUSE r, AND ∠A = θ (THETA—IT'S TRADITIONAL!).

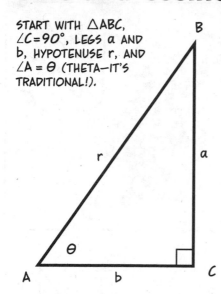

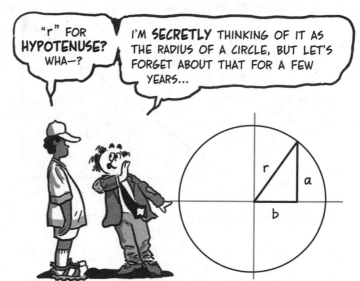

"r" FOR **HYPOTENUSE**? WHA—?

I'M **SECRETLY** THINKING OF IT AS THE RADIUS OF A CIRCLE, BUT LET'S FORGET ABOUT THAT FOR A FEW YEARS...

Definitions

THE **SINE** OF θ, WRITTEN sin θ, IS THE RATIO a/r.

$$\sin \theta = \frac{\text{SIDE OPPOSITE } \theta}{\text{HYPOTENUSE}}$$

THE **COSINE** OF θ, WRITTEN cos θ, IS THE RATIO b/r.

$$\cos \theta = \frac{\text{SIDE ADJACENT } \theta}{\text{HYPOTENUSE}}$$

MATHEMATICIANS HAVE FOUND WAYS TO CALCULATE THESE RATIOS TO VERY HIGH ACCURACY. MY PHONE'S CALCULATOR GIVES THEM TO **15** DECIMAL PLACES. FOR THE SINE, FIRST PUNCH IN THE ANGLE MEASURE, THEN THE "SIN" KEY. FOR THE COSINE, HIT "COS."

I NEVER KNEW WHAT THAT KEY WAS, BUT I'VE ALWAYS BEEN TEMPTED!

$\sin 25° = 0.422618261740699...$

$\cos 89° = 0.017452406437283...$

$\sin 30° = 0.5 \ (=\frac{1}{2})$

$\cos 45° = \frac{\sqrt{2}}{2} = 0.707106781186547...$

Examples

1. IN A 3-4-5 RIGHT TRIANGLE

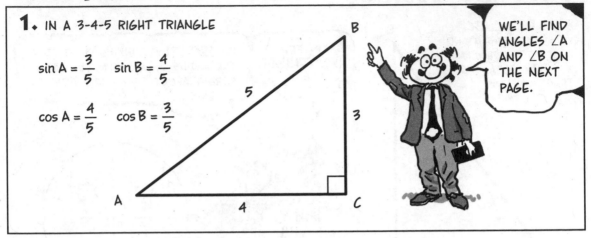

$$\sin A = \frac{3}{5} \qquad \sin B = \frac{4}{5}$$

$$\cos A = \frac{4}{5} \qquad \cos B = \frac{3}{5}$$

WE'LL FIND ANGLES ∠A AND ∠B ON THE NEXT PAGE.

2. THIS RIGHT TRIANGLE HAS ONE KNOWN ANGLE, ∠A=54°, AND ONE KNOWN SIDE, b=5. BY DEFINITION,

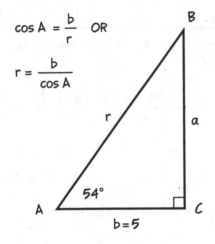

$$\cos A = \frac{b}{r} \qquad OR$$

$$r = \frac{b}{\cos A}$$

WE FIND cos 54° ON THE CALCULATOR

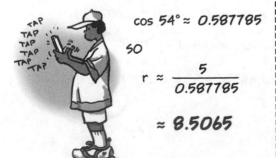

cos 54° ≈ 0.587785

SO

$$r \approx \frac{5}{0.587785}$$

$$\approx 8.5065$$

NOW WE CAN FIND SIDE a BECAUSE

$$\sin 54° = \frac{a}{r} \qquad OR$$

$$a = r \sin 54°$$

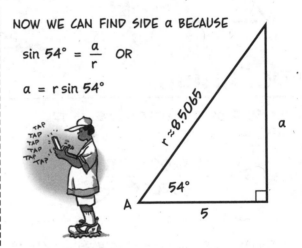

AGAIN USE THE CALCULATOR TO FIND sin 54° ≈ 0.809017 AND PLUG THAT IN:

$$a \approx (8.5065)(0.809017)$$

$$\approx 6.8819$$

TRIANGLE SOLVED!

216

IN RIGHT TRIANGLES, WE CAN ALSO CALCULATE ANGLES BASED ON SIDES.

IN A 3-4-5 RIGHT TRIANGLE, WE FOUND

$$\sin \angle A = \frac{3}{5} = 0.6$$

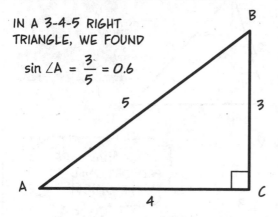

AND THEN? HOW DO WE FIND $\angle A$ FROM ITS SINE? AGAIN, BREAK OUT THE CALCULATOR.

1. ENTER 0.6.

2. PRESS THE "INV" OR "2ND FUNCTION" KEY.

3. PRESS $\sin^{-1}$.

4. READ **36.8698...**

$\angle A = $ **36.8698...°**

 THIS $\sin^{-1}$ IS CALLED THE **INVERSE SINE** OR **ARCSINE**. FEED IT A NUMBER BETWEEN ZERO AND 1, AND IT RETURNS **THE ANGLE WITH THAT SINE.** IN THIS CASE, THE ANGLE HAVING SINE 0.6 IS AROUND 36.87°. THERE IS ALSO AN INVERSE COSINE KEY, $\cos^{-1}$.

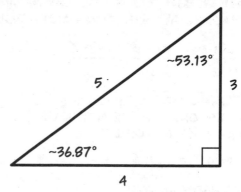

$\angle B$, OF COURSE, IS THE COMPLEMENT OF 36.87°.

$$\angle B \approx 90° - 36.87$$
$$\approx \mathbf{53.13°}$$

TRIANGLE SOLVED!

HOW DID ANYONE EVER, UM, FUNCTION BEFORE CALCULATORS?

WITH GIANT TABLES!

WE SOLVE OTHER TRIANGLES BY DIVIDING THEM INTO TWO RIGHT TRIANGLES. IF THE ALTITUDE BD = h, THEN

$$\frac{h}{a} = \sin C \qquad h = a\sin C$$

$$\frac{h}{c} = \sin A \qquad h = c\sin A$$

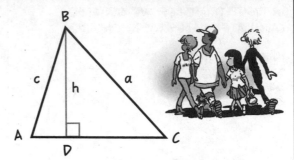

Theorem 20-1 (THE LAW OF SINES). IN A
TRIANGLE △ABC WITH THREE ACUTE ANGLES

$$\frac{\sin A}{a} = \frac{\sin B}{b} = \frac{\sin C}{c}$$

Proof. WE HAVE TWO DIFFERENT
EXPRESSIONS FOR THE ALTITUDE h, WHICH MUST BE EQUAL.

$$c\sin A = a\sin C$$

DIVIDING BY ac GIVES

$$\frac{\sin A}{a} = \frac{\sin C}{c}$$

SIMILARLY, USING THE ALTITUDE FROM A IN THE SAME WAY GIVES

$$\frac{\sin C}{c} = \frac{\sin B}{b} \ \blacksquare$$

THE **SIDES** ARE PROPORTIONAL TO THE **SINES**!!

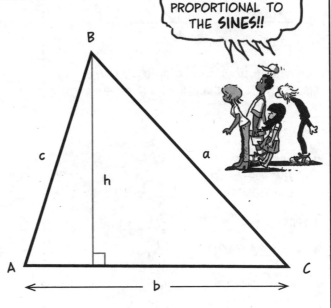

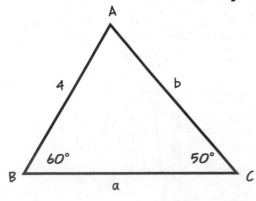

Example: SUPPOSE WE KNOW TWO ANGLES AND A SIDE.

PLUGGING IN GIVES

$$\frac{\sin C}{c} \approx \frac{\sin 50°}{4} \approx \frac{0.766}{4} \approx 0.1915$$

BY THE LAW OF SINES,

$$0.1915 \approx \frac{\sin B}{b} = \frac{\sin 60°}{b} \approx \frac{0.86}{b}, \text{ SO}$$

$$b \approx \frac{0.86}{0.1915} \approx \textbf{4.491}$$

SINCE ∠A = 70°, WE CAN FIND a ≈ 4.907 IN THE SAME WAY.

Theorem 20-2 (THE LAW OF COSINES). IF TRIANGLE △ABC HAS THREE ACUTE ANGLES AND SIDES a, b, AND c, OPPOSITE VERTICES A, B, AND C, RESPECTIVELY, THEN

$$c^2 = a^2 + b^2 - 2ab \cos C$$

Proof. AN ALTITUDE AD = h DIVIDES THE BASE CB INTO TWO PARTS CD = p AND DB = q, WITH p + q = a.

$$\cos C = \frac{p}{b}, \quad SO \quad p = b \cos C$$

BY PYTHAGORAS,

$$b^2 - p^2 = h^2 = c^2 - q^2$$

SUBSTITUTING a − p FOR q,

$$b^2 - p^2 = c^2 - (a - p)^2$$

$$b^2 - p^2 = c^2 - a^2 + 2ap - p^2$$

$$c^2 = a^2 + b^2 - 2ap$$

$$= a^2 + b^2 - 2ab \cos C \ \blacksquare$$

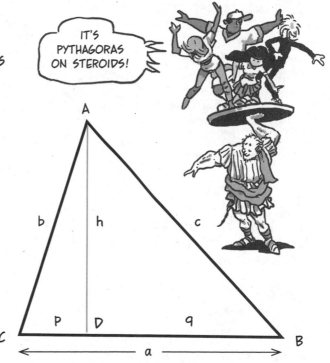

IT'S PYTHAGORAS ON STEROIDS!

Example: SUPPOSE a = 10, b = 11, ∠C = 34°. BY THE LAW OF COSINES,

$$c^2 = 100 + 121 - 2(10)(11) \cos 34° \approx 221 - (220)(0.82904)$$

$$\approx 221 - 182.39 = 38.61$$

$$c \approx \sqrt{38.61} \approx \mathbf{6.213}$$

BY THE LAW OF SINES,

$$\frac{\sin A}{10} \approx \frac{\sin C}{6.213}$$

$$\sin A \approx \frac{(10)(0.5592)}{6.213} \approx 0.900$$

$$\angle A \approx \sin^{-1}(0.900) \approx \mathbf{64.12°}$$

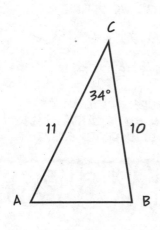

TRIANGLE SOLVED!

219

WITH THESE TWO LAWS WE CAN SOLVE ANY TRIANGLE—OR RATHER, ANY TRIANGLE WITHOUT AN OBTUSE ANGLE. SO FAR, SINES AND COSINES ARE DEFINED ONLY FOR ANGLES LESS THAN 90°, EVEN THOUGH SOME TRIANGLES HAVE OBTUSE ANGLES.

CAN WE MAKE SENSE OF sin θ AND cos θ HERE? MATHEMATICIANS DO IT, AND THEIR REASONING MAY SOUND A LITTLE WEASELLY, AT FIRST...

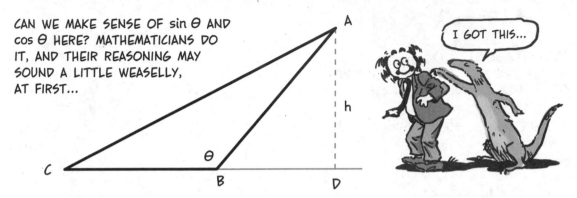

I GOT THIS...

IT GOES LIKE THIS: THE POINT A IS STILL **ABOVE** THE BASE BC, SO THE **SINE** SHOULD STILL BE **POSITIVE.** WE DEFINE IT AS

MAKES PERFECT SENSE TO ME...

$$\sin \theta = \frac{AD}{AB}$$

BUT THE POINT D IS ON THE **OPPOSITE** SIDE OF B FROM THE ANGLE, SO THE **COSINE** SHOULD BE **NEGATIVE.**

$$\cos \theta = \frac{-BD}{AB}$$

AND I'M A WEASEL WITH A PhD!!

SO WE MAKE THIS A **DEFINITION:** IF $90° < \theta < 180°$, WE DEFINE

$$\sin \theta = \sin(180° - \theta)$$

$$\cos \theta = -\cos(180° - \theta)$$

WITH THESE VALUES, THE LAWS OF SINES AND COSINES ARE **UNIVERSALLY TRUE** AND CAN BE USED TO SOLVE

ALL TRIANGLES.

THE SINE AND COSINE ARE CALLED **TRIGONOMETRIC** FUNCTIONS BECAUSE THEY SOLVE THE TRIANGLE-MEASUREMENT PROBLEM, BUT THEY ALSO APPEAR IN SUBJECTS FAR FROM GEOMETRY, SUCH AS MUSIC AND ANYTHING ELSE HAVING TO DO WITH WAVES AND VIBRATION.

$$\sum_{j=1}^{\infty} a_j \sin jx + b_j \cos jx,$$

AMIRIGHT?

THAT'S ONE WAY TO ENJOY A CONCERT...

YOU'LL HAVE TO GO A LITTLE FARTHER IN MATH, THOUGH, TO UNDERSTAND WHY. IN THIS BOOK (OR WHAT'S LEFT OF IT) WE'LL STICK TO USING "TRIG" FUNCTIONS ON THE SUBJECT AT HAND, AND YOU KNOW WHAT THAT IS...

221

Exercises USE A CALCULATOR, UNLESS IT SAYS OTHERWISE.

1. WHAT IS sin 10°? sin 20°? sin 80°? sin 90°? DO SINES GET LARGER AS ANGLES INCREASE? WHY DO YOU THINK THAT sin 90° HAS THIS VALUE? WHAT IS cos 80°? cos 70°?

2. FIND THE UNKNOWN QUANTITIES.

a.
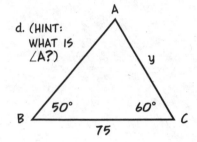

b.
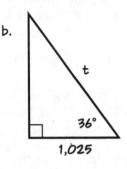

c.
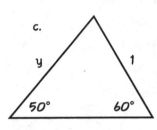

d. (HINT: WHAT IS ∠A?)
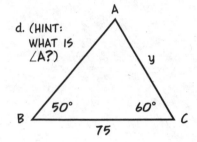

e.
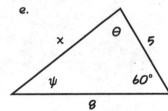

f.
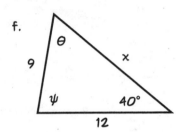

3. ON P. 214, WE SAW cos 45° = $\frac{1}{2}\sqrt{2}$.

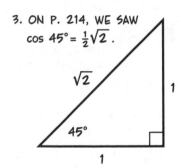

NOW CONSIDER △ABC WITH AB=AC=1 AND ∠A=45°.

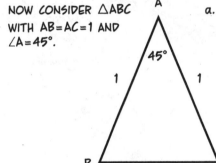

a. USE THE LAW OF COSINES TO FIND THE BASE BC.

b. SHOW THAT sin 22$\frac{1}{2}$° = $\frac{1}{2}\sqrt{2-\sqrt{2}}$.

c. CONFIRM IT WITH YOUR CALCULATOR.

4.
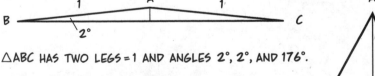

△ABC HAS TWO LEGS =1 AND ANGLES 2°, 2°, AND 176°.

a. THE ALTITUDE OF THE TRIANGLE = sin 2°. FIND THIS ON A CALCULATOR.

b. USE PYTHAGORAS TO FIND BC.

c. BY THE LAW OF SINES, WHAT IS sin 176°?

d. HOW CAN sin 176° BE SO SMALL WHEN THE ANGLE IS SO BIG?

5. IN △ABC, DRAW ALTITUDES AD AND BC. SHOW THAT

(BC)(EC) = (AC)(DC).

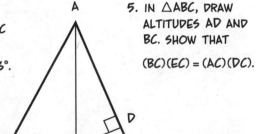

Chapter 21
POLYGONS
MORE SIDES THAN A BARREL OF BEES

A POLYGON IS JUST LIKE A TRIANGLE OR QUADRILATERAL, ONLY MORE SO: MORE SIDES, MORE ANGLES, MORE VERTICES.

YES, WE'RE INDIVIDUALS, BUT WE'RE ALL THE SAME **KIND** OF INDIVIDUAL!

BEES MAKE SIX-SIDED **HEXAGONS**; THE U.S. MILITARY HAS A FIVE-SIDED **PENTAGON**; STOP SIGNS ARE EIGHT-SIDED **OCTAGONS**... AND WE ALSO SAY THINGS LIKE "15-GON" WHEN THE NUMBER OF SIDES IS LARGE (OR "N-GON" WHEN WE WANT TO BE VAGUE).

AND NOW—LET'S BEE-GONE!

MORE PRECISELY:

Definition. A POLYGON WITH N SIDES IS A SEQUENCE OF N POINTS, CALLED THE VERTICES, $V_1, V_2, \ldots, V_N$, NO TWO POINTS THE SAME, AND THE SEGMENTS OR SIDES, $V_1V_2, V_2V_3, \ldots, V_NV_{N-1}$, THAT JOIN THEM IN ORDER, PLUS A FINAL SEGMENT V_NV_1 THAT CLOSES THE CHAIN.

IN ADDITION, NO SIDE MAY INTERSECT ANOTHER, ASIDE FROM THE SHARED ENDPOINTS.

NEVER THAT!

THIS!

Postulate 12. EVERY POLYGON DIVIDES THE PLANE INTO AN INTERIOR AND EXTERIOR REGION WITH NO POINTS IN COMMON. THE POLYGON IS THE BOUNDARY OR PERIMETER OF THE INTERIOR REGION.

FROM NOW ON, WE CONSIDER ONLY CONVEX POLYGONS, MEANING THAT IF TWO POINTS ARE IN THE INTERIOR, THEN SO IS THE ENTIRE SEGMENT BETWEEN THEM.

IN A CONVEX POLYGON, A **DIAGONAL**—A SEGMENT BETWEEN ANY TWO NONADJACENT VERTICES—LIES INSIDE THE POLYGON. FROM ANY VERTEX, **N-3** DIAGONALS CAN BE DRAWN. (A DIAGONAL GOES TO ALL BUT THREE VERTICES: THE POINT ITSELF AND ITS TWO NEIGHBORS.)

THIS FORMS N-2 TRIANGLES... AND THE POLYGON'S INTERIOR ANGLES HAVE THE SAME TOTAL AS THE ANGLES OF ALL THESE TRIANGLES PUT TOGETHER, SO...

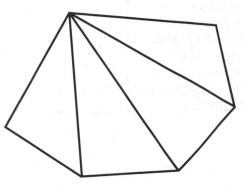

SIX SIDES, FOUR TRIANGLES!

Theorem 21-1. THE INTERIOR ANGLES OF A POLYGON WITH N SIDES SUM TO

$(N-2)\,180°.$ ▮

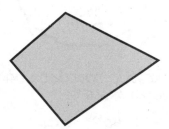

$2 \times 180° = \mathbf{360°}$

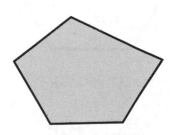

$3 \times 180° = \mathbf{540°}$

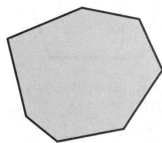

$5 \times 180° = \mathbf{900°}$

FROM THIS WE INFER THE TOTAL OF A POLYGON'S **EXTERIOR** ANGLES, TOO. EACH EXTERIOR ANGLE IS SUPPLEMENTARY TO AN INTERIOR ANGLE. IF **S** IS THE SUM OF THE EXTERIOR ANGLES, THEN

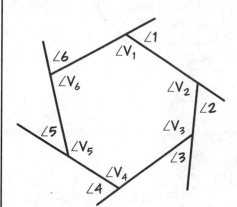

$S = \angle 1 + \angle 2 + \ldots + \angle N$

$= (180° - \angle V_1) + (180° - \angle V_2) + \ldots + (180° - \angle V_N)$

$= N \cdot 180° - (\angle V_1 + \angle V_2 + \ldots + \angle V_N)$

$= N \cdot 180° - (N-2)180°$

$= 2 \times 180° = \mathbf{360°}$

NO MATTER HOW MANY SIDES.

Definition. A POLYGON IS **REGULAR** IFF ALL ITS SIDES ARE EQUAL AND ALL ITS ANGLES ARE EQUAL.

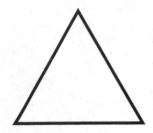

REGULAR 3-GONS ARE **EQUILATERAL TRIANGLES.** REGULAR QUADRILATERALS ARE **SQUARES.**

EACH INTERIOR ANGLE AT ANY VERTEX V OF A REGULAR POLYGON IS 1/N TIMES THE SUM OF ALL THE ANGLES.

$$\angle V = \frac{(N-2)(180°)}{N}$$

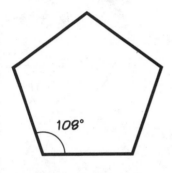

108°

120°

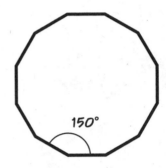

150°

A REGULAR PENTAGON'S VERTEX ANGLE IS

$$\frac{3(180°)}{5} = \mathbf{108°}$$

A REGULAR HEXAGON'S IS

$$\frac{4(180°)}{6} = \mathbf{120°}$$

A REGULAR DODECAGON (N=12) HAS AN ANGLE OF

$$\frac{10(180°)}{12} = \mathbf{150°}$$

A REGULAR 100-GON'S INTERIOR ANGLE IS

$$\frac{98(180°)}{100} = \mathbf{176.4°}$$

IS THAT EVEN AN ANGLE?

WHY DON'T WE MAKE ONE LIKE THAT?

WE'RE BUSY, NOT INSANE...

AS THE LAST EXAMPLE SHOWS, A REGULAR POLYGON WITH A GREAT MANY SIDES IS SOMETHING LIKE AN "APPROXIMATE CIRCLE."

SOME-THING LIKE A WHAT?

AN EXPRES-SION WITH AN APPROXIMATE MEANING.

IN FACT, A REGULAR POLYGON WITH JUST A FEW SIDES HAS SOMETHING CIRCULAR ABOUT IT. FOR ONE THING, IT APPEARS TO HAVE A SORT OF **CENTER** EQUIDISTANT FROM EVERY VERTEX (IF NOT FROM OTHER POINTS ON THE PERIMETER).

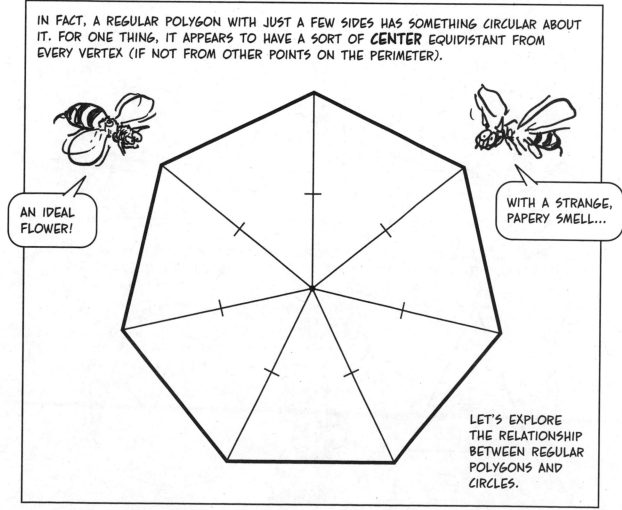

AN IDEAL FLOWER!

WITH A STRANGE, PAPERY SMELL...

LET'S EXPLORE THE RELATIONSHIP BETWEEN REGULAR POLYGONS AND CIRCLES.

Circles Around Polygons

Definition. A POLYGON IS **CYCLIC** IF ALL ITS VERTICES LIE ON SOME CIRCLE. THE CYCLIC POLYGON IS SAID TO BE **INSCRIBED** IN THE CIRCLE; THE CIRCLE **CIRCUMSCRIBES** THE POLYGON.

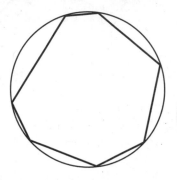

IS THIS A PROBLEM?

YOU CAN MAKE A CYCLIC POLYGON BY CONNECTING ANY FINITE COLLECTION OF POINTS ON A CIRCLE IN ORDER.

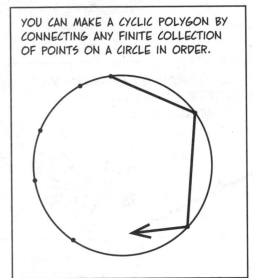

BUT NOT ALL POLYGONS ARE CYCLIC. HERE A CIRCLE THROUGH ANY THREE POINTS CAN'T POSSIBLY PASS THROUGH THE FOURTH ONE.

NO WAY... JUST... NO... WAY...

SAD!

ALL TRIANGLES ARE CYCLIC. AS WE'VE SEEN, THE PERPENDICULAR BISECTORS OF ANY TWO SIDES INTERSECT AT THE CENTER OF THE CIRCUMSCRIBING CIRCLE.

THREE POINTS DETERMINE A CIRCLE, REMEMBER?

228

Theorem 21-2. EVERY **REGULAR** POLYGON IS CYCLIC. GIVEN A POLYGON $V_1V_2...V_N$ WITH EQUAL SIDES ($V_1V_2 = V_2V_3 = ... = V_NV_1$) AND EQUAL ANGLES $\angle V_1 = \angle V_2 = \angle V_3 = ... \angle V_N$, A CIRCUMSCRIBING CIRCLE CAN BE CONSTRUCTED.

START WITH THREE POINTS, AND THE REST FOLLOW...

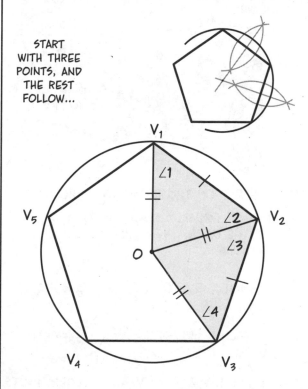

Proof. WE ILLUSTRATE WITH A PENTAGON, BUT THE ARGUMENT IS THE SAME FOR ANY NUMBER OF SIDES.

1. FIND THE CENTER O OF THE CIRCLE PASSING THROUGH THE **THREE POINTS** V_1, V_2, AND V_3 AS ABOVE. $OV_1 = OV_2 = OV_3$, SO $\triangle OV_1V_2$ AND $\triangle OV_2V_3$ ARE ISOSCELES.

2. $V_1V_2 = V_2V_3$ BY ASSUMPTION.

3. $\triangle OV_1V_2 \cong \triangle OV_2V_3$ BY SSS.

4. $\angle 1 = \angle 3$, $\angle 2 = \angle 4$ (CORR. PARTS)

5. $\angle 1 = \angle 2$, $\angle 3 = \angle 4$ (ISOSCELES)

6. $\angle 1 = \angle 2 = \angle 3 = \angle 4 = \frac{1}{2}\angle V_1V_2V_3$

NOW WE SHOW THAT $OV_4 = OV_3$.

7. $\angle V_2V_3V_4 = \angle V_1V_2V_3$ (ASSUMED)

8. $\angle 4 = \frac{1}{2}\angle V_2V_3V_4$ (SUBST.)

9. $\angle 5 = \angle 4$ (SUBTRACTION)

10. $\angle 5 = \angle 3$ (SUBST.)

11. $\triangle OV_3V_4 \cong \triangle OV_2V_3$ (SAS)

12. $OV_4 = OV_3$ (CORR. PARTS) THAT IS, **V_4 IS ON THE CIRCLE.**

13. REPEATING THE ARGUMENT ONE VERTEX AT A TIME SHOWS THAT ALL THE REMAINING VERTICES ARE ON THE CIRCLE. ∎

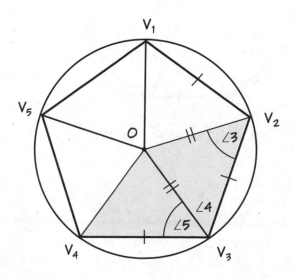

Fitting Polygons into Circles

GIVEN A REGULAR POLYGON, IT'S EASY TO CIRCUMSCRIBE A CIRCLE: JUST PASS IT THROUGH THREE VERTICES. GIVEN A CIRCLE, CAN WE INSCRIBE A REGULAR POLYGON? THE ANSWER IS:

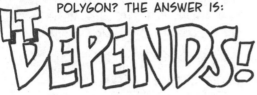

WHEN THEY ARE POSSIBLE, CONSTRUCTIONS OFTEN DEPEND ON THIS REASONABLE THEOREM.

Theorem 21-3.
IF A CIRCLE HAS A NUMBER OF EQUALLY SPACED RADII, SUCH THAT THE ANGLE BETWEEN EACH ADJACENT PAIR IS THE SAME, THEN THE POLYGON FORMED BY THEIR ENDPOINTS ON THE CIRCLE IS REGULAR.

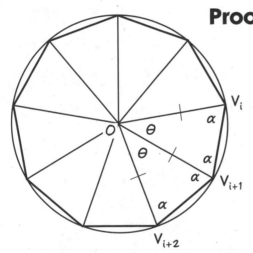

$\angle V_i V_{i+1} V_{i+2} = 2\alpha$ FOR EACH i

Proof.

1. FOR EACH i, $OV_i = OV_{i+1}$, $\angle V_i OV_{i+1} = \angle V_{i+1} OV_{i+2} = \theta$.   (ASSUMED)

2. FOR EACH VALUE OF i, $\triangle V_i OV_{i+1} \cong \triangle V_{i=1} OV_{i+2}$.   (SAS)

3. FOR ALL i, $V_i V_{i+1} = V_{i+1} V_{i+2}$.   (CORR. PARTS)

4. FOR EACH i, $\triangle V_i OV_{i+1}$ HAS EQUAL BASE ANGLES $\alpha = (\frac{1}{2}(180° - \theta))$.   ($\triangle V_i OV_{i+1}$ ISOSCELES)

5. FOR EACH i, $\angle V_i V_{i+1} V_{i+2} = 2\alpha$.   (ADDITION)

6. WITH ALL SIDES EQUAL AND ALL ANGLES EQUAL, THE POLYGON IS REGULAR. ∎   (DEF. OF REGULAR)

Equilateral Triangle

GIVEN A CIRCLE CENTERED AT O, DRAW A DIAMETER COP.

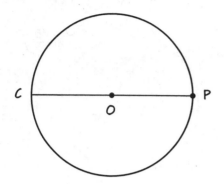

DRAW A SECOND CIRCLE OF RADIUS r CENTERED AT P.

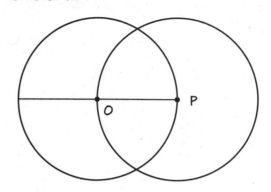

IF A AND B ARE THE POINTS WHERE THE CIRCLES INTERSECT, THEN △ABC IS EQUILATERAL.

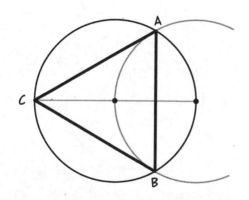

WHY? △AOP AND △BOP ARE EQUILATERAL, SO ∠AOP = ∠BOP = 60°. SO ∠AOC = ∠COB = 180°– 60° = 120° = ∠AOB. THE THREE CENTRAL ANGLES BEING EQUAL, △ABC IS REGULAR BY THEOREM 21-3.

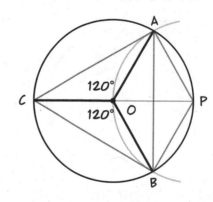

Square

ALMOST AS EASY: DRAW TWO PERPENDICULAR DIAMETERS AND CONNECT THEIR ENDPOINTS. THE POLYGON IS REGULAR BECAUSE ALL CENTRAL ANGLES EQUAL 90°.

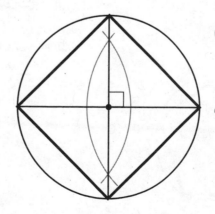

OH, FOR HEAVEN'S SAKE! IT'S STILL SQUARE!

231

Pentagon
THESE CONSTRUCTIONS DEPEND ON THE MATERIAL IN CHAPTER 19 ABOUT THE GOLDEN RATIO $\phi = \frac{1}{2}(1 + \sqrt{5})$ AND THE 36°-72°-72° TRIANGLE.

GIVEN A CIRCLE OF RADIUS r CENTERED AT O, DRAW TWO PERPENDICULAR RADII OA AND OP.

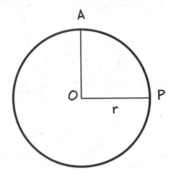

FIND THE MIDPOINT M OF OA. DRAW MP.

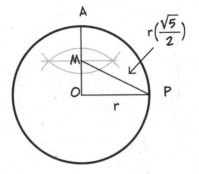

$r\left(\frac{\sqrt{5}}{2}\right)$

CENTERING COMPASS AT M, EXTEND OA TO OQ WITH MQ = MP.

$OQ = \frac{r}{2} + \frac{r\sqrt{5}}{2}$

$= r\phi$

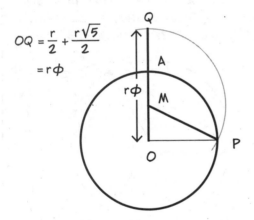

CENTERING COMPASS AT Q, DRAW ARC OF RADIUS OQ = ϕ INTERSECTING CIRCLE AT B AND E. DRAW AB AND AE.

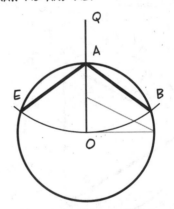

MARK POINTS C AND D ON CIRCLE WITH BC = AB, DE = AE AND COMPLETE THE FIGURE.

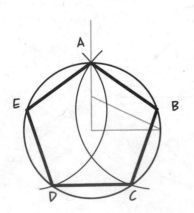

THIS WORKS BECAUSE △BQO AND △OQE HAVE LEGS rϕ AND BASE r, MAKING ∠QOB=72°, THE REGULAR PENTAGON'S CENTRAL ANGLE.

∠BOA = ∠AOE = 72°

$= \frac{360°}{5}$

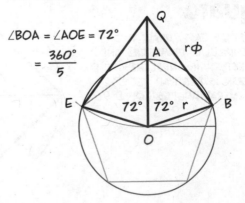

232

ANOTHER CONSTRUCTION USES THE FACT (SHOWN IN THE EXERCISES) THAT A REGULAR PENTAGON OF RADIUS r HAS SIDE s EQUAL TO

$$s = r\left(\sqrt{1 + \frac{1}{\phi^2}}\right)$$

DRAW A DIAMETER PQ AND A PERPENDICULAR RADIUS OA. BISECT OQ AT M. DRAW AM.

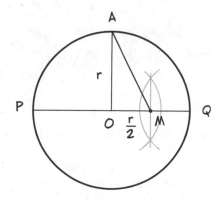

CENTERING COMPASS AT M, MARK S ON PQ WITH MS = MA = $r\sqrt{5}/2$.

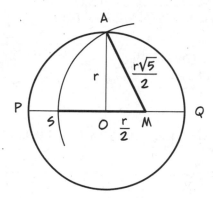

THEN

$$OS = r\frac{\sqrt{5}-1}{2} = r(\phi - 1) = \frac{r}{\phi}$$

(BECAUSE $\phi - 1 = \frac{1}{\phi}$),

SO BY PYTHAGORAS,

$$AS^2 = r^2 + \frac{r^2}{\phi^2}$$

$$AS = r\left(\sqrt{1 + \frac{1}{\phi^2}}\right)$$

THE DESIRED SIDE LENGTH.

YEESH!

THEY GET EASIER AFTER THIS, PROMISE!

CENTERING COMPASS AT A, DRAW ARC OF RADIUS AS TO FIND B AND E.

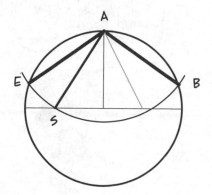

FIND C AND D AS BEFORE AND COMPLETE THE FIGURE.

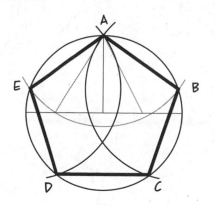

Hexagon

ANOTHER EASY ONE, BECAUSE A REGULAR HEXAGON HAS ITS SIDE EQUAL TO ITS RADIUS. THE POLYGON CONSISTS OF SIX EQUILATERAL TRIANGLES!

$$\text{CENTRAL ANGLE} = \frac{360°}{6} = 60°$$

THAT'S WHY THEY FIT TOGETHER SO NICELY!

THE SIMPLEST CONSTRUCTION DUPLICATES THE STEPS FOR AN EQUILATERAL TRIANGLE.

AT EACH END OF A DIAMETER, CENTER TWO ARCS WITH THE SAME RADIUS AS THE CIRCLE. THE HEXAGON'S VERTICES ARE THE DIAMETER'S ENDPOINTS PLUS THE INTERSECTIONS OF THE CIRCLE AND ARCS.

Heptagon

THE 7-GON IS EITHER EASIER OR HARDER, DEPENDING ON YOUR POINT OF VIEW. IT SIMPLY CAN'T BE DONE! DON'T EVEN TRY!

I'LL TAKE THAT AS A CHALLENGE...

SEE YOU IN A FEW HUNDRED YEARS, THEN...

OTHER STRAIGHTEDGE-AND-COMPASS IMPOSSIBILITIES HAVE NINE, ELEVEN, AND FIFTEEN SIDES. AROUND THE YEAR 1800, THE ÜBER-WHIZ **GAUSS** (SEE P. 96) CONSTRUCTED A **17**-GON AND SHOWED THAT IN THEORY A **65,537**-GON WAS POSSIBLE, TOO.

FIRST GET A REALLY, **REALLY** BIG SHEET OF PAPER!

ON THE OTHER HAND, WE CAN ALWAYS **DOUBLE** THE NUMBER OF SIDES OF A GIVEN REGULAR POLYGON BY BISECTING ITS CENTRAL ANGLES OR ITS SIDES. A SQUARE (N = 4) LEADS TO AN OCTAGON (N = 8); A PENTAGON (N = 5) TO A DECAGON (N = 10); ETC.

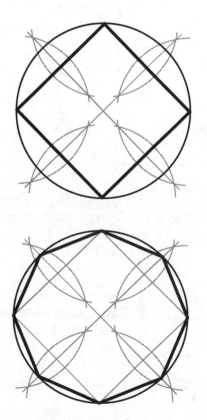

AND THAT, READERS, IS ALL WE HAVE TO SAY ON THIS SUBJECT.

WHAT? NOTHING ABOUT PRIME NUMBERS N OF THE FORM—

SHOOSH! **ABSOLUTELY NOTHING!**

Polygons as Near Circles

IN ALL THE TALK ABOUT
LENGTH AND AREA IN THIS
BOOK, WE'VE NEVER MEN-
TIONED CIRCLES. WHAT'S
THE **LENGTH** OF AN **ARC**?

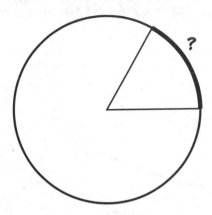

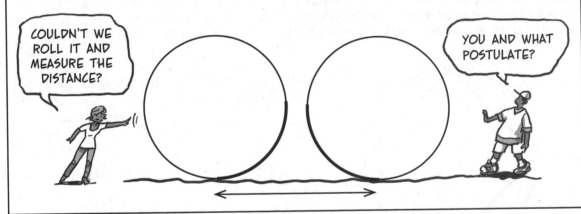

THE QUESTION REMAINS UNANSWERED BECAUSE THE RULER POSTULATE APPLIES ONLY
TO STRAIGHT LINES. OUR RULERS ARE STRAIGHT AND RIGID, NOT FLOPPY LIKE TAPE
MEASURES. TO US, THE LENGTH OF A CURVE HAS NO MEANING—YET.

COULDN'T WE
ROLL IT AND
MEASURE THE
DISTANCE?

YOU AND WHAT
POSTULATE?

POLYGONS, ON THE
OTHER HAND, HAVE
STRAIGHT SIDES. WE
CAN DIVIDE THEM
INTO TRIANGLES AND
USE ALL THE TOOLS
AT OUR DISPOSAL,
INCLUDING THE TRIG
FUNCTIONS.

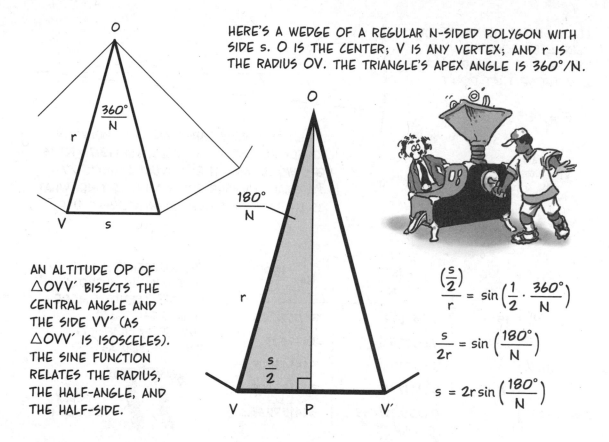

HERE'S A WEDGE OF A REGULAR N-SIDED POLYGON WITH SIDE s. O IS THE CENTER; V IS ANY VERTEX; AND r IS THE RADIUS OV. THE TRIANGLE'S APEX ANGLE IS 360°/N.

$\dfrac{360°}{N}$

r

V s

$\dfrac{180°}{N}$

r

$\dfrac{s}{2}$

V P V'

AN ALTITUDE OP OF △OVV′ BISECTS THE CENTRAL ANGLE AND THE SIDE VV′ (AS △OVV′ IS ISOSCELES). THE SINE FUNCTION RELATES THE RADIUS, THE HALF-ANGLE, AND THE HALF-SIDE.

$$\frac{\left(\frac{s}{2}\right)}{r} = \sin\left(\frac{1}{2} \cdot \frac{360°}{N}\right)$$

$$\frac{s}{2r} = \sin\left(\frac{180°}{N}\right)$$

$$s = 2r\sin\left(\frac{180°}{N}\right)$$

THE POLYGON'S **PERIMETER**—THE DISTANCE P ALL THE WAY AROUND, OR THE SUM OF ALL THE SIDES—IS N × s, AS ALL THE SIDES ARE THE SAME LENGTH.

$$P = N \times 2r\sin\left(\frac{180°}{N}\right)$$

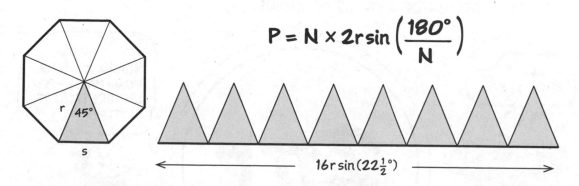

r $45°$

s

$16r\sin\left(22\frac{1}{2}°\right)$

FOR INSTANCE, IN A REGULAR OCTAGON, θ=45° (360°/8), θ/2 = 22.5°, AND FROM LAST CHAPTER'S EXERCISES,

$$\sin 22.5° = \left(\frac{\sqrt{2-\sqrt{2}}}{2}\right)$$

SO ONE SIDE IS

$$2r \cdot \frac{1}{2}\left(\sqrt{2-\sqrt{2}}\right) \approx r(0.7653686)$$

AND THE PERIMETER IS APPROXIMATELY

$$8r(0.7653686) = \mathbf{6.122949r}$$

LET'S GROUP TERMS OF THE PERIMETER FORMULA INTO THOSE THAT DEPEND ON N AND THOSE THAT DON'T.

$$P = (2r)\left(N\sin\frac{180°}{N}\right)$$

TERMS DEPENDING ON N

WHEN N IS VERY LARGE AND THE POLYGON CLOSELY RESEMBLES A CIRCLE, $\sin(180°/N)$ IS VERY SMALL, AND THE PRODUCT $N\sin(180°/N)$ IS... A SMALL NUMBER TIMES A LARGE ONE. WHAT IS IT? LET'S PUT OUR CALCULATOR TO USE.

N	$\dfrac{180°}{N}$	$\sin\dfrac{180°}{N}$	$N\sin\dfrac{180°}{N}$
12	15°	0.258819...	3.1058285...
50	3.6°	0.06279...	3.139525...
180	1°	0.01745...	3.14143315...
1,800	0.1°	0.001745...	3.14159105...
180,000	0.001°	0.000017453...	3.14159265...

THE NUMBERS IN THE LAST COLUMN APPROACH A **LIMIT** SLIGHTLY MORE THAN 3.14159265, A NUMBER FAMILIAR TO MANY OF YOU AS

GREEK LETTER PI

AND WE CONCLUDE, BY SOME MYSTERIOUS PROCESS OF "PASSING TO THE LIMIT," THAT C, **THE CIRCUMFERENCE OF A CIRCLE OF RADIUS r,** IS

HEY, I KNEW THAT!

WELL... YOU'VE **HEARD** IT... AND NOW—MAYBE—YOU **UNDERSTAND** IT BETTER!

2πr

238

WHAT ABOUT **AREA?** EACH TRIANGULAR WEDGE HAS AREA $\mathcal{A}_w = \frac{1}{2}hs$, WHERE h IS AN ALTITUDE. BUT

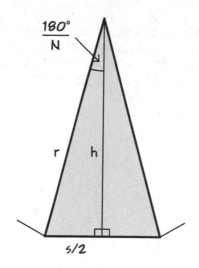

$$\frac{s}{2} = r\sin\frac{180°}{N} \qquad h = r\cos\frac{180°}{N}$$

SO

$$\mathcal{A}_w = \frac{hs}{2} = r^2\sin\frac{180°}{N}\cos\frac{180°}{N}$$

AND THE POLYGON'S **TOTAL AREA** $\mathcal{A}$ IS N OF THESE:

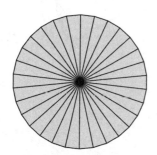

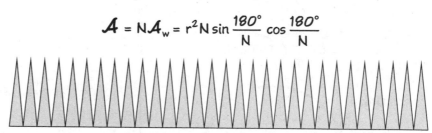

$$\mathcal{A} = N\mathcal{A}_w = r^2 N\sin\frac{180°}{N}\cos\frac{180°}{N}$$

WHEN N IS VERY LARGE, $N\sin(180°/N)$ APPROACHES π, AS WE SAW. MEANWHILE $\cos(180°/N)$ NEARS 1 BECAUSE h BECOMES "ALL BUT EQUAL TO" r. IN SHORT—

$$\mathcal{A} = r^2 \underbrace{N\sin\frac{180°}{N}}_{\downarrow \atop \pi} \underbrace{\cos\frac{180°}{N}}_{\downarrow \atop 1}$$

$$\mathcal{A} = \pi r^2$$

AND I THOUGHT THAT FORMULA DROPPED FROM HEAVEN...

HEY, WHERE DO YOU THINK THE TRIG FUNCTIONS COME FROM?

NOW WE HAVE A REASONABLE WAY TO TALK ABOUT THE LENGTH OF A CIRCULAR ARC: THE ARC TAKES UP SOME FRACTION OF THE WHOLE CIRCLE, AS MEASURED BY THE ARC'S **ANGLE**.

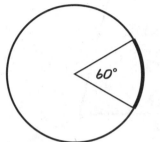

ARC = 60°/360° OR 1/6 OF THE CIRCLE.

ITS LENGTH, THEN, SHOULD BE THAT SAME FRACTION OF THE TOTAL CIRCUMFERENCE, 2πr.

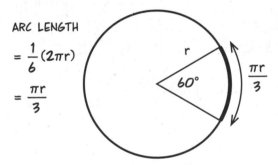

ARC LENGTH

$= \dfrac{1}{6}(2\pi r)$

$= \dfrac{\pi r}{3}$

THIS ALSO WORKS FOR ARCS >180°: THIS ARC HAS A DEGREE MEASURE GIVEN BY THE REFLEX ANGLE THAT SPANS IT.

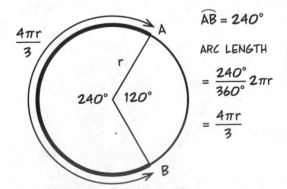

$\widehat{AB} = 240°$

ARC LENGTH

$= \dfrac{240°}{360°}2\pi r$

$= \dfrac{4\pi r}{3}$

IN GENERAL, SUPPOSE θ IS THE ANGULAR MEASURE IN DEGREES OF AN ARC $\widehat{AB}$.

Definition. THE **ARC LENGTH** OF $\widehat{AB}$ IS 2πrθ/360°, OR πrθ/180°.

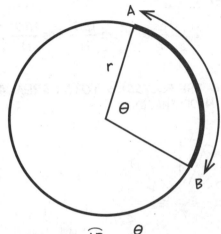

LENGTH $\widehat{AB} = \dfrac{\theta}{180°}\pi r$

TO MEASURE ARC LENGTH, THERE'S NO NEED FOR A BENT RULER. AN ORDINARY RULER (FOR THE RADIUS) AND A PROTRACTOR (FOR THE ANGLE) WILL DO.

SO YOU CAN TOSS THAT, ER, THING, THEN...

AFTER ALL THIS—TOSS?

THIS BRINGS OUR CHAPTER ON POLYGONS TO A POSSIBLY SURPRISING END...

WE STARTED WITH BASIC, STRAIGHT-SIDED FIGURES AND LEARNED SOMETHING ABOUT THEIR ANGLES, INSIDE AND OUT.

AROUND A **REGULAR** POLYGON, AT LEAST, WE SAW THAT CIRCUMSCRIBING A CIRCLE WAS EASY, WHILE INSCRIBING A REGULAR POLYGON IN A CIRCLE WAS TRICKIER, OR EVEN IMPOSSIBLE.

SIGH...

THEN WE SHIFTED GEARS! WE UPPED THE NUMBER OF SIDES—A **LOT**—AND, VOILÀ, A NUMBER π APPEARED AS A LIMITING VALUE OF $N\sin(180°/N)$ AS N "WENT TO INFINITY."

AND THAT, LADIES, GENTLEMEN, AND OTHERS, IS YOUR FORETASTE OF **CALCULUS!**

WHICH MEANS IT'S TIME TO STOP—FOR NOW!

Exercises

1. △ABC IS INSCRIBED IN A CIRCLE OF RADIUS 1.

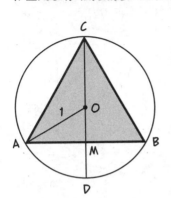

a. WHAT IS ∠OAB?

b. WHAT IS THE LENGTH OF OM? OF MD?

c. HOW LONG IS SIDE AB?

d. WHAT IS THE ALTITUDE OF △ABC?

e. WHAT IS THE AREA OF △ABC?

f. WHAT IS THE AREA OF AN INSCRIBED HEXAGON?

2. MOMO HAS A RECIPE CALLING FOR A 9-INCH CIRCULAR PAN. ALL SHE HAS IS A 9×9 SQUARE PAN. BY WHAT PERCENTAGE DOES SHE NEED TO INCREASE THE RECIPE?

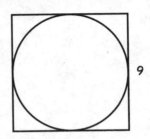

3. QUADRILATERAL ABCD IS INSCRIBED IN A CIRCLE. ∠B SPANS AN ARC $\overset{\frown}{AC}$.

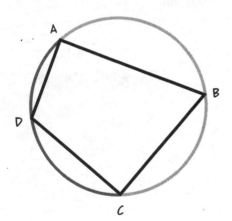

a. WHAT ARC DOES ∠D SPAN?

   i. THE SAME ARC AS ∠B

   ii. AN ARC $\overset{\frown}{CA}$ THAT CONSISTS OF EVERY-THING IN THE CIRCLE **EXCEPT** $\overset{\frown}{AC}$

b. IN DEGREES, WHAT IS $\overset{\frown}{AC} + \overset{\frown}{CA}$?

c. WHAT IS ∠B+∠D?

d. WHAT IS ∠A+∠C?

e. WHAT THEOREM HAVE YOU JUST PROVED ABOUT CYCLIC QUADRILATERALS? DO YOU THINK THE CONVERSE IS TRUE? WHY OR WHY NOT?

4. IN THAT PESKY 36-72-72 TRIANGLE, SUPPOSE THE BASE=1, SO THE LEGS = $\phi = \frac{1}{2}(1+\sqrt{5})$. USE THE LAW OF COSINES TO FIND THE BASE IN TERMS OF THE SIDES AND THE APEX ANGLE.

a. USE THE EQUATION TO SHOW

$$\cos 36° = 1 - \frac{1}{2\phi^2}$$

b. BY THE MAGIC OF ϕ, $1/\phi^2 = 1 - (1/\phi)$ AND $1/\phi = \phi - 1$. CONCLUDE THAT

$$\cos 36° = \frac{\phi}{2}$$

c. VERIFY THIS WITH A CALCULATOR.

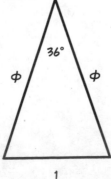

d. IN A 72-54-54 TRIANGLE WITH LEG=1, WHY IS h=ϕ/2?

e. USE THE PYTHAGOREAN THEOREM TO SHOW THAT

$$BC = \sqrt{1 + \frac{1}{\phi^2}}$$

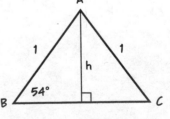

[HINT: $3 - \phi = 1 + 1 - (\phi-1) = 1 + (1/\phi^2)$]

f. THIS IS THE SIDE OF A REGULAR PENTAGON OF RADIUS 1. WHY?

SOLUTIONS TO SELECTED PROBLEMS

Solutions to Selected Problems

Chapter 1, p. 12

1. $c^2 = 20 + 9 = 29$, SO $c = \sqrt{29}$.   2. YES, PLEASE.

Chapter 2, p. 28

1a. WE PROBABLY CAN'T IMAGINE CURVED 3-D SPACES BECAUSE WE HAVE NO EXPERIENCE OF THEM.

1b. ONE IDEA: ON A CURVED SURFACE, ENLARGING A FIGURE CAN CHANGE ITS SHAPE, SO MAYBE SOMETHING LIKE THAT WOULD ALSO HAPPEN IN A CURVED 3-D SPACE.

2. POSTULATE 2. IT SAYS THAT ANY POINT P ALWAYS "SEES" ANOTHER POINT A, AND POSTULATE 1 SAYS THAT A LINE CONTAINS A AND P.

3. YES.

4. ONLY ONCE. IF TWO LINES SHARE MORE THAN A SINGLE POINT, THEY'RE THE SAME LINE, BY POSTULATE 1.

5c. HYPOTHESIS: KLEPTO-MART CHARGES TWO DOLLARS FOR A BAG OF CHIPS. CONCLUSION: YOU CAN GET A BAG OF CHIPS FOR LESS THAN TWO DOLLARS SOMEWHERE BESIDES KLEPTO-MART.

6. CONVERSE OF 5c: IF YOU CAN GET A BAG OF CHIPS FOR LESS THAN TWO DOLLARS SOMEWHERE BESIDES KLEPTO-MART, THEN KLEPTO-MART CHARGES TWO DOLLARS FOR A BAG OF CHIPS.

CONTRAPOSITIVE OF 5c: IF YOU CAN'T GET A BAG OF CHIPS FOR LESS THAN TWO DOLLARS SOMEWHERE BESIDES KLEPTO-MART, THEN KLEPTO-MART DOESN'T CHARGE TW O DOLLARS FOR A BAG OF CHIPS.

7. A DOG LIVES IN BROOKLYN **OR** A DOG HAS SPOTS.

Chapter 3, p. 36

2. $AB = 1 - (-4) = 1 + 4 = 5$, $BC = 3\frac{1}{2} - 1 = 2\frac{1}{2}$

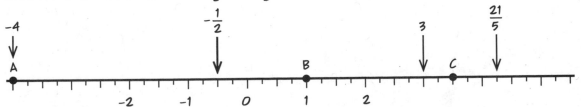

3. THE DISTANCE, 5.125, DOESN'T DEPEND ON THE NUMBERING. EVERY "LOWER" COORDINATE IS TWO MORE THAN THE CORRESPONDING "UPPER" COORDINATE: IF P HAS UPPER COORDINATE p AND LOWER COORDINATE p′, THEN p′ = p+2. IF Q IS ANY OTHER POINT ON THE LINE, $PQ = |p'-q'| = |p+2-(q+2)| = |p-q|$. THE TWOS CANCEL.

4b. $-3 - (-5) = 2$ 4d. 2   5a. YES.   5b. NO.   5c. YES.   5d. NO.   6. YES.

7. FIRST NOTE THAT AB>0 AND CD>0. THEN

AD = AB + BC + CD, SO AD - BC = AB + CD > 0

8. $b > a \Rightarrow b - a > a - a \Rightarrow b - a > 0$

Chapter 4, p. 46

1. TO TRIPLE AB, FIRST DOUBLE IT: EXTEND THE SEGMENT, CENTER THE COMPASS AT B, AND DRAW A CIRCLE OF RADIUS AB, CROSSING THE LINE AT C. ADD ANOTHER LENGTH AB WITH A CIRCLE OF THE SAME RADIUS CENTERED AT C.

TO QUADRUPLE AB, DOUBLE AC TO AE. TO MULTIPLY AB BY 8, DOUBLE AE.

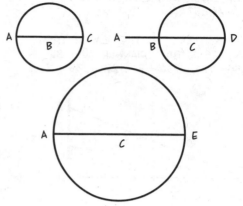

2. RADIUS IS $CQ = PC - PQ = 88.8 - 55.3 = 33.5$.

3. BECAUSE $AB + BC = AC$.   4. NO.

5a. $r_B > AB + r_A$ OR $r_B - r_A > AB$

5c. UPPER: $r_B - r_A = AB$ LOWER: $r_B + r_A = AB$

Chapter 5, p. 58

1c. OBTUSE. 1d. ACUTE.   2. 140°   4. 360°/3 = 120°   5. 172°   7. $\angle DVE = 27°$, SO $\angle CVE = 117°$.

8. THE DARK SHADED ANGLE IS ONE.

9. $\angle DCG = \angle ACB$ (VERTICAL ANGLES)
   $\angle ACB = \angle FGH$ (GIVEN)
   $\angle FGH + \angle CGF = \angle DCG + \angle CGF = 180°$ (LINEAR PAIR)

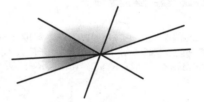

Chapter 6, p. 74

1a. YES. 1b. NO.   1c. ABSOLUTELY NOT!   1d. YES, BY SSS.

2. THE VERTICAL ANGLES AT C ARE EQUAL, SO THE TRIANGLES ARE CONGRUENT BY SAS.

3. EITHER R OR R′ MAKES A SIDE EQUAL TO AC.

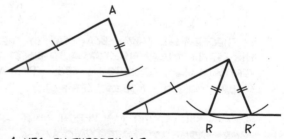

4. YES, BY THEOREM 6-3.

5. A↔V, B↔T, C↔U

Chapter 7, p. 82

1. THEY ARE VERTICAL ANGLES.   2. 115°

3a. THE EXTERNAL ANGLE AT B IS 53°,
SO ∠A<53°.

3b. ∠C<53° ALSO, SO THE SUM OF THE
ANGLES IN THIS TRIANGLE MUST BE
LESS THAN 53°+53°+127°=233°.

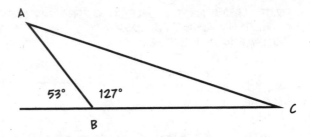

5a. BY THEOREM 7-3, THE PERPENDICULAR IS THE
SHORTEST SEGMENT JOINING THE POINT TO THE LINE.

5b. △ODB≅△OCB BY SAS. SO ∠ODB=90°.

5c. AB>BD BY THEOREM 7-3. BD=BC, SO AB>BC.

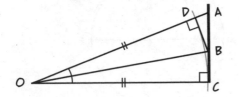

6c AND 6d ARE IMPOSSIBLE TRIANGLES.   6f. THE LEGS MUST BE >4.

Chapter 8, p. 90

3. AFTER RAISING THE PERPENDICULAR AND FIND-
ING THE SEGMENT'S MIDPOINT M, USE A COM-
PASS TO MARK POINT R SO THAT QR=MQ.

4. THE SIDES ARE RADII OF EQUAL CIRCLES.

6. DRAW ARCS OF RADIUS DC CENTERED AT A
AND B. THEIR INTERSECTION IS THE APEX
OF THE TRIANGLE.

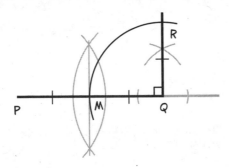

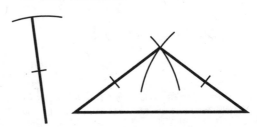

7. BY THEOREM 6-1, ∠ABC=∠BCA=∠CAB. BY THEOREM 6-3,
THE PERPENDICULAR BISECTORS ALSO BISECT THE
VERTEX ANGLES, SO ∠ABQ=½∠ABC=∠PCB. AB=BC BY
ASSUMPTION, SO △AQB≅△CPB BY ASA.

ALSO AQ=BR=½AC BY ASSUMPTION. SINCE
∠BMR=∠AQM=90°, ∠MBR=∠MAQ BY ASA.

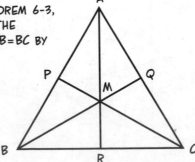

Chapter 9, p. 98

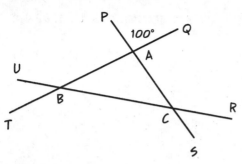

1a. GREATER.   1b. 80°   1c. LESS.   1d. LESS.

2. NO, CAN'T PROVE ANY OF THEM.

3c. NO.   4e IS CORRECT.

Chapter 10, p. 108

1a. YES, AT AN ANGLE OF 1°.   1b. 114°   1c. ∠1   1d. 63°

2. AB∥DC, SO ∠BDC=∠ABD. AD∥BC, SO ∠CBD=∠ADB. BD=BD. THE CONGRUENCE FOLLOWS FROM ASA.

3a. 105°   3c. 58°

4a. START WITH ∠2+∠5+∠6=180° AND
∠5+∠9=180°=∠6+∠10.

4b. HERE ARE SOME ANGLES FILLED IN.

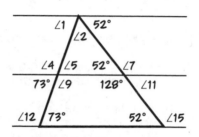

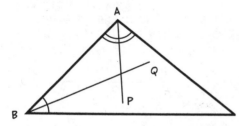

5. AB IS TRANSVERSE TO AP AND BQ. ADJACENT INTERIOR ANGLES ∠ABP AND ∠BAP ARE HALF OF TWO ANGLES OF A TRIANGLE, SO ∠ABQ+∠BAP<90°, SO YEAH, AP INTERSECTS BQ!!!

6. ∠CAD<∠CAB, ∠C=∠C. SUBTRACT THEM FROM 180° FOR THE RESULT.

Chapter 11, p. 116

2a. 80°   2b. 180°–114°=66°   3a. 110° AND 30°.   4. 360°   5. AAS (VERTICAL ANGLES ARE EQUAL).

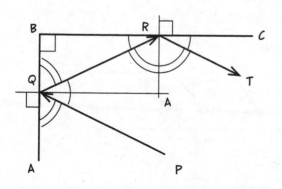

6. DRAW RA AND QA PERPENDICULAR TO THE MIRRORS AT POINTS Q, R. BRAQ IS A RECTANGLE, SO ∠RQB=∠QRA AND ∠RQA=∠QRA, AND ∠QRA+∠QRB=90°.

BY THE LAW OF REFLECTION, ∠ARQ=∠ART AND ∠AQR=∠AQP.

THUS ∠CRT=∠QRB (BOTH = 90°–∠ARQ).

THEN ∠PQR+∠QRT = ∠QRB+∠CRT+∠QRT = 180°.

SO RT∥PQ.

Chapter 12, p. 130

1a. 68°   1b. 360°−192°=168°   1c. 69°   2a. YES.   2b. NO, BECAUSE 59+119=178, NOT 180.

3. 360°−125°−125° = 110°   4a. SAS.

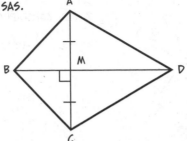

5. ALL ARE 90°.

4b. ALSO △AMD ≅ △CMD, SO AD=CD AND △ABD ≅ CBD BY SSS.

6. YES.

Chapter 13, p. 146

1a. 4   1c. 30   1d. 6   1e. $\frac{15}{2}$ 2a. AB=DC = $\frac{22.548}{5.637}$ = 4   3. HALF, OR 2,338,002.

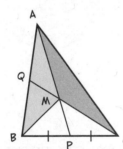

4a. △APB AND △APC HAVE EQUAL BASES AND ALTITUDES.

4b. MP IS A MEDIAN OF THE WHITE TRIANGLE, SO △BMP AND PMC HAVE EQUAL AREAS. SUBTRACT THESE FROM THE RESULT OF 4a.

4c. A RESULT OF 4a AND 4b. AND YES, EXTENDING BM MAKES ANOTHER MEDIAN, SO THAT △ABC IS DIVIDED INTO SIX TRIANGLES OF EQUAL AREA. M IS CALLED THE **CENTER OF GRAVITY** OF THE TRIANGLE.

5. AM= MC SO △APM ≅ △CQM BY ASA. (WHICH ANGLES?) THEN A(APQD)= A(△ADC) (WHY?), WHICH IS HALF THE AREA OF THE PARALLELOGRAM BECAUSE △ADC ≅ △CBA.

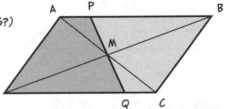

6. 2

Chapter 14, p. 156

1a. 29   1b. 15   1d. 28   2a. $\sqrt{5}$ 2b. $\sqrt{2}$ 2c. 2

3. AB=BD (WHY?); DE=DC (WHY?); ∠ADC=∠ADE (WHY?)

5. THE ALTITUDE IS 4, BY PYTHAGORAS. THE AREA IS 32, SO THE BASE IS 32/4=8.

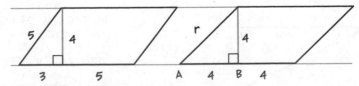

IN THE "TILTED" FIGURE, THEN, AB=4 AND r = $\sqrt{16+16}$ = $4\sqrt{2}$.

Chapter 15, p. 174

1a. $10x=40$, $x=4$ 1b. $x^2=81$, $x=9$ 2a. YES. 2b. NO. 3. $\dfrac{30}{20}=\dfrac{9}{x}$, $x=6$

4. THIS IS A WAY TO MAKE PARALLEL LINES AT ANY SPACING. TO MAKE LINES $(3/7)''$ APART, TILT THE RULER SO THAT 7 OF ITS INCHES SPAN 3 VERTICAL INCHES.

5a. BY SSAS. 5d. YES. 6a. $\dfrac{AP}{AB}=\dfrac{AS}{AD}=\dfrac{1}{2}$

6b–e. ALL YES, BY SIDE-SPLITTER THEOREM.

6f. A PARALLELOGRAM.

7. IT'S NOT NEEDED: TWO ANGLES (AA) ARE ENOUGH!

Chapter 16, p. 182

1. d IS CORRECT.

2. IT HELPS TO ENHANCE THE DIAGRAM AS AT RIGHT TO SEE WHY THE RESULTS ARE TRUE.

2a. $\dfrac{1}{4}$ 2b. $\dfrac{1}{3}$

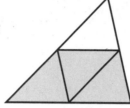

5. FIRST SHOW THAT

$\dfrac{c^2}{ab}=\dfrac{f^2}{de}$. THEN $\dfrac{c^2}{ab}-\dfrac{1}{2}=\dfrac{f^2}{cd}-\dfrac{1}{2}$

SOME ALGEBRA TURNS THIS INTO

$\dfrac{c^2-\frac{1}{2}ab}{ab}=\dfrac{f^2-\frac{1}{2}cd}{cd}$ OR $\dfrac{\mathcal{A}}{ab}=\dfrac{\mathcal{A}'}{de}$

3. AS IN 2, TRY DIVIDING $\triangle ABC$ INTO SMALLER CONGRUENT TRIANGLES. OF COURSE YOU CAN ALSO SOLVE IT BY USING PROPORTIONS.

4. IF THE ORIGINAL SQUARE HAS SIDE s, THEN THE SMALL SHADED SQUARE HAS SIDE $\dfrac{s}{2\sqrt{2}}$ AND AREA $s^2/8$.

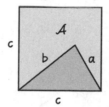

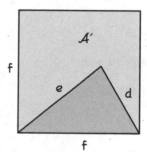

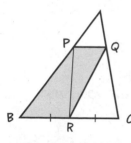

6. $\dfrac{3}{5}$. DRAW PR. $\triangle PBR$ AND $\triangle QRC$ HAVE EQUAL BASES AND ALTITUDES, HENCE EQUAL AREAS. $\triangle PQR$ HAS THE SAME ALTITUDE AND A BASE $2/3$ AS LONG. SO

$$\mathcal{A}(PQRB)=\mathcal{A}(\triangle QRC)+\tfrac{2}{3}\mathcal{A}(\triangle QRC)=\tfrac{5}{3}\mathcal{A}(\triangle QRC)$$

Chapter 17, p. 194

1a. $5x=66$, $x=13\frac{1}{5}$ 1c. $x^2=9\times16=144$, $x=12$

2a. $\triangle PBC \sim \triangle PDA$

2b. BECAUSE $\dfrac{PA}{PC}=\dfrac{PD}{PB}$.

2c. BECAUSE $\angle PBC=180°-\angle ABC$, $\angle PDA=180°-\angle ADC$.

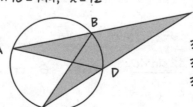

3a. BY AAS.

3b. BECAUSE $AB\perp PQ$ AND $AC\perp PR$.

3c. BECAUSE D IS ON BOTH ANGLE BISECTORS.

3d. YES.

Chapter 18, p. 204

1a. $3.125 \times 2 = 6.25$, $\sqrt{6.25} = 2.5$ 1c. $\sqrt{20} = 2\sqrt{5}$ 1d. $s\sqrt{t}$

2. THE GEOMETRIC MEAN OF ANY NUMBER AND ZERO IS ZERO; THE GEOMETRIC MEAN OF ANY NUMBER n AND 1 IS $\sqrt{n}$.   3a. 5g   3b. rg   3c. 4   5. YES.

6. ONE WAY TO SQUARE A TRIANGLE: FIRST DOUBLE THE TRIANGLE TO FORM A RECTANGLE; THEN SQUARE THE RECTANGLE; AND FINALLY JOIN THE MIDPOINTS OF THE SQUARE TO MAKE A SQUARE WITH HALF THE AREA. (ANOTHER: HALVE THE RECTANGLE AND SQUARE THAT.)

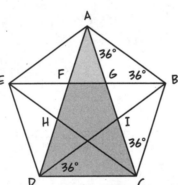
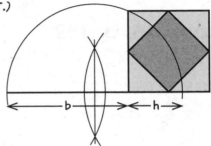

Chapter 19, p. 212

1a. $\phi + 1 = \phi^2$, SO $\dfrac{\phi+1}{\phi} = \dfrac{\phi^2}{\phi} = \phi$ 1b. $\dfrac{AE}{EF} = \phi$ 1c. $\phi - 1 = \dfrac{1}{\phi}$, SO $\dfrac{1}{\phi-1} = \phi$

1d. $\phi^{15} + \phi^{14} = \phi^{14}(\phi+1) = \phi^{14}\phi^2 = \phi^{16}$ 2. $BD = 1$, $CD = 1/\phi$, $CE = CD = 1/\phi$, $ED = 1/\phi^2$

3a. FILLING IN SOME ANGLES SHOWS THAT $EB \parallel DC$. SO $\triangle AFG \sim \triangle ADC$.

3b. THE "POINT" TRIANGLES ARE CONGRUENT, SO $AG = CI$, AND

$$\frac{CG}{AG} = \phi$$

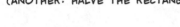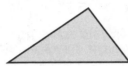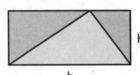

3b (CONT'D). THEN

$$\frac{AC}{AG} = \frac{CG+AG}{AG} = \frac{CG}{AG} + 1 = \phi + 1 = \phi^2$$

3d. ϕ^4

4. ϕs

DOESN'T ϕ JUST DRIVE YOU NUTS?

Chapter 20, p. 222

2a. $x = 3\sin 20° \approx 1.026$ 2b. $t = 1{,}025\cos 36° \approx 829.24$ 2c. $y = \dfrac{\sin 60°}{\sin 50°} \approx 1.1305$

2e. $x^2 = 5^2 + 8^2 - (2)(5)(8)\cos 60° = 89 - 40 = 49$, $x = \sqrt{49} = 7$

2f. $\dfrac{\sin \theta}{12} = \dfrac{\sin 40°}{9}$, SO $\sin \theta = \dfrac{12\sin 40°}{9} \approx 0.857$, $\theta \approx \arcsin 0.857 = 58.99°$

3a. BY THE LAW OF COSINES, $BC^2 = 1 + 1 - 2\cos 45° = 2 - \sqrt{2}$, SO $BC = \sqrt{2 - \sqrt{2}}$. THEN $\sin 22\frac{1}{2}° = \frac{1}{2}BC = \frac{1}{2}\sqrt{2 - \sqrt{2}}$.

4. ABOVE 90°, THE SINE GETS SMALLER AS THE ANGLE GETS LARGER.

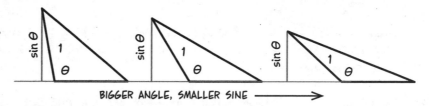

BIGGER ANGLE, SMALLER SINE ⟶

Chapter 21, p. 242

1a. 30°   1b. $OM = MD = \frac{1}{2}$

1c. $AM = \cos 30° = \frac{1}{2}\sqrt{3}$, $AB = \sqrt{3}$

1d. $\frac{3}{2}$ 1e. $\frac{3}{4}\sqrt{3}$

1f. $\frac{3}{2}\sqrt{3}$, TWICE THE TRIANGLE, AS THE HEXAGON ADDS THREE TRI-ANGLES CONGRUENT TO $\triangle AOC$.

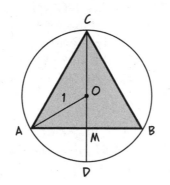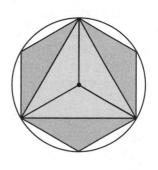

2. $\mathcal{A}(\text{CIRCLE}) = \frac{\pi}{4}(9)^2$, $\mathcal{A}(\text{SQUARE}) = 9^2$. THE RATIO $= 4/\pi \approx 1.27$. INCREASE THE RECIPE BY 27%.

3. IN A CYCLIC QUADRILATERAL, OPPOSITE ANGLES ARE SUPPLEMENTARY. THE CONVERSE IS ALSO TRUE: IF OPPOSITE ANGLES ADD TO 180°, THE QUADRILATERAL IS CYCLIC. TRY PROVING IT BY DRAWING A CIRCLE THROUGH THREE VERTICES AND REASONING FROM THERE.

4a-b. BY THE LAW OF COSINES,

$$1 = \phi^2 + \phi^2 - 2\phi^2\cos 36° \quad \text{SO}$$

$$\frac{1}{\phi^2} = 1 + 1 - 2\cos 36°$$

$$2\cos 36° = 2 - \frac{1}{\phi^2} = 1 + \left(1 - \frac{1}{\phi^2}\right)$$

$$= 1 + \frac{1}{\phi} = \phi \quad \text{SO}$$

$$\cos 36° = \frac{\phi}{2}$$

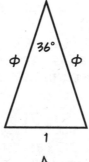

4d. BECAUSE $h = \cos 36°$.

4e. BY PYTHAGORAS,

$$\frac{BC^2}{4} + \frac{\phi^2}{4} = 1, \quad \frac{BC^2}{4} + \frac{\phi + 1}{4} = 1$$

$$BC^2 = 3 - \phi = 2 - (\phi - 1) = 2 - \frac{1}{\phi}$$

$$= 1 + 1 - \frac{1}{\phi} = 1 + \frac{1}{\phi^2}$$

4f. THIS IS THE SIDE OF A REGULAR PENTAGON OF RADIUS 1 BECAUSE THE PENTAGON'S CENTRAL ANGLE IS 72° (TWICE 36°).

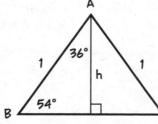

Acknowledgments

ACKNOWLEDGMENTS ARE LIKE OSCARS ACCEPTANCE SPEECHES. WHO PAYS ANY ATTENTION TO THEM, ASIDE FROM THE PEOPLE BEING NAMED, AND THE PEOPLE NOT BEING NAMED? BUT HERE GOES, ANYWAY...

MR. PHILLIPS TAUGHT ME HIGH SCHOOL ALGEBRA BACK IN PHOENIX, ALTHOUGH (EVEN AFTER YEARS OF GRAD SCHOOL!) I NEVER UNDERSTOOD THE PARALLEL POSTULATE PROPERLY UNTIL THIS BOOK WAS WELL UNDERWAY. FOR THAT I OWE A DEBT OF GRATITUDE TO DAVID JOYCE OF CLARK UNIVERSITY, WHO PUT AN ANNOTATED EDITION OF EUCLID'S *ELEMENTS* ONLINE AT https://mathcs.clarku.edu/~djoyce/elements/bookI/bookI.html. AND LET'S NOT FORGET TIM BERNERS-LEE, WHO STARTED THE WORLDWIDE WEB. THERE'S A LOT OF GOOD MATH ON THE WEB, NOT LEAST ON WIKIPEDIA. JIMMY WALES, YOU ROCK!

MY WIFE, LISA GOLDSCHMID, ENDURED MY SORRY FLIRTATION WITH RETIREMENT AFTER I FINISHED *THE CARTOON GUIDE TO BIOLOGY*, UNTIL WE REALIZED THAT WE'D BOTH BE HAPPIER IF I HAD A PROJECT TO WORK ON.

MY AGENT, VICKY BIJUR, DID THE DEAL AND ALWAYS GIVES UNQUALIFIED ENCOURAGEMENT. MY EDITOR, NICK AMPHLETT, IS ANOTHER BOTTOMLESS SOURCE OF GOOD VIBES, PLUS HE RUNS VALUABLE INTERFERENCE WITH THE PESKY PROOFREADERS.

APPLE COMPUTER MADE THE IPAD ON WHICH MUCH OF THE FINAL ART WAS DRAWN, A FIRST FOR ME. SAVAGE INTERACTIVE MADE PROCREATE, THE GREAT PAINTING AND DRAWING APP I USED. EVENTUALLY IT ALL ENDED UP IN ADOBE ILLUSTRATOR, PHOTOSHOP, AND INDESIGN. THANKS, JOHN WARNOCK, FOR FOUNDING ADOBE.

ZENO JOHNSON REALLY DID TELL US TO DEFINE OUR TERMS, PROBABLY GOOD ADVICE FOR WRITING ESSAYS, ADVICE THAT IS BEING TEDIOUSLY AND SCRUPULOUSLY FOLLOWED BY THE LARGE LANGUAGE MODELS CURRENTLY BEGINNING TO SPEW PROSE FROM THEIR SUPERSIZED NEURAL NETS.

FINALLY, IF YOU'RE AMONG THE SEVEN BILLION PEOPLE I FAILED TO MENTION, I'M SURE THAT YOU HAD SOME EFFECT ON THE PROCEEDINGS, LARGE OR SMALL, SO THANKS TO YOU, TOO. I'M SURE YOU UNDERSTAND THAT I CAN'T NAME YOU ALL INDIVIDUALLY BECAUSE WE'RE RUNNING OUT OF PAPER. MAYBE IN THE EBOOK?

INDEX

Index

AAS (ANGLE-ANGLE-SIDE; THEOREM 11-3), 111

ACUTE ANGLES, 52
 EXERCISES, 58
 TRIANGLES, 79, 214, 218–19

ADJACENT ANGLES, 55, 57, 60, 76, 106

ALTERNATING INTERIOR ANGLES, 106

ALTITUDE, 136, 145
 TRAPEZOIDS AND PARALLELOGRAMS, 142–43
 TRIANGLES, 136–40, 153, 169, 176–77, 218

ANGLES, 47–58
 BISECTING, 70, 85
 IN CIRCLES, 49–50, 186–88
 COMPLEMENTARY, 57, 110
 CONGRUENT, 61–66
 COPYING, 84
 CORRESPONDING, 93–95, 99, 101–3, 106
 DEFINITION, 49
 DEGREES, 4, 50–51
 EXERCISES, 58, 245
 EXTERIOR. SEE EXTERIOR ANGLES
 INSCRIBED, 187–89
 OBTUSE, 52, 58, 220
 PERPENDICULAR, 56, 69–71
 POSTULATE 7 (PROTRACTOR POSTULATE), 53–54
 PROTRACTORS, 51–54
 REFLEX, 240
 RIGHT, 56, 79, 110, 112–13
 SUPPLEMENTARY, 57, 106
 TERMINOLOGY AND NOTATION, 57

THEOREM 5-1, 54–55
THEOREM 5-2, 55
THEOREM 5-3, 56
THEOREM 15-2, 167–68
TRIANGLES, 23
VERTEX OF THE, 49, 52–53, 55, 60
VERTICAL, 55–56, 82, 94, 106

ANGLE BISECTORS, 70, 85–86, 144, 206
 EXERCISES, 90, 194

ANGLE-SIDE-ANGLE (POSTULATE 9), 65–66, 73

APEX, 68–70, 205

ARCS, 57, 85–86, 89, 186
 EXERCISES, 90

AREA, 131–46
 DEFINITION, 133
 EXERCISES, 146, 248
 POSTULATE 11, 134
 PYTHAGOREAN THEOREM, 149–51
 QUADRILATERALS, 141–42
 RECTANGLES, 132–34
 SCALING. SEE SCALING
 THEOREM 13-1, 135
 THEOREM 13-2, 136–37
 THEOREM 13-3, 139–40
 THEOREM 13-4, 142
 THEOREM 14-1, 148
 TRAPEZOIDS AND PARALLELOGRAMS, 142–44
 TRIANGLES, 135–40

ARMY ANTS, 47–48

ASA (ANGLE-SIDE-ANGLE; POSTULATE 9), 65–66, 73

ASPECT RATIOS, 160, 162, 182

AVERAGE
    OF THE BASES, 142
    GEOMETRIC. SEE GEOMETRIC MEAN

BABYLONIANS, 4–5, 50

BIRKHOFF, GEORGE, 27

BISECTORS. SEE ANGLE BISECTORS;
    PERPENDICULAR BISECTORS

BISECTING ANGLES, 70, 85

CARROLL, LEWIS, 117

CARTOON GUIDE TO ALGEBRA, THE
    (GONICK), 29

CENTRAL ANGLES, 127, 186–89

CHORDS, 185, 187–89
    EXERCISES, 194, 249
    GEOMETRIC MEAN, 195–98

CIRCLES, 37–46, 183–94
    ANGLES IN, 49–50, 186–88
    AROUND POLYGONS, 228–29
    CHORDS, 185, 187–89
    CIRCUMSCRIBING, 228–30
    COMPASS, 9, 37–39
    DEFINITION, 39
    DIAMETER, 39, 51
    DRAWING, 37–38
    EXERCISES, 46, 194, 245, 249
    FITTING POLYGONS INTO, 230–31
    GEOMETRIC MEAN, 195–200
    PIE SLICES, 4
    POLYGONS AS NEAR, 236–40
    POSTULATE 6, 43–44
    RADIUS, 39–41, 43–44, 184, 238
    TANGENTS, 190–93
    THEOREM 4-1, 40–41
    THEOREM 4-2, 42

THEOREM 17-1, 185
THEOREM 17-2, 188
THEOREM 17-3, 190–91, 193
THEOREM 17-4, 191
THROUGH THREE NONCOLLINEAR
    POINTS, 87–89, 105

CIRCULAR DEFINITIONS, 14

CIRCUMFERENCE, 238, 240

CLASSICAL CONSTRUCTIONS, 83–90
    BISECTING ANGLES, 85
    CIRCLE THROUGH THREE NONCOLLINEAR
    POINTS, 87–89, 105
    COPYING ANGLES, 84
    EXERCISES, 90, 246
    PERPENDICULAR BISECTOR OF
    SEGMENTS, 85
    PERPENDICULAR THROUGH POINTS, 86
    THEOREM 8-1, 88–89

COLLINEAR POINTS, 20, 31–32

COMMON FACTOR, 202

COMPASS, 9, 37–39
    THEOREM 4-1, 40–41

COMPLEMENTARY ANGLES, 57, 110

COMPLEMENTS, 25

CONCLUSION, 21

CONGRUENCE, 61–69, 155
    EXERCISES, 74, 82, 130

CONGRUENCE TESTS, 64–66
    ASA, 65–66, 73
    QUADRILATERALS, 125, 127
    RIGHT TRIANGLES, 113
    SAS, 65–66, 73
    SSS, 72–73

CONSTRUCTION INDUSTRY, USE OF
    GEOMETRY, 2

CONSTRUCTIONS, 44, 165–69

CONTRAPOSITIVES, 25, 28, 95, 104

CONVERSE STATEMENTS, 22–23, 28

CONVEX FIGURES, 118

CONVEX POLYGONS, 224–25

COORDINATES, 30–32, 40
    EXERCISES, 36, 46, 244, 245

COPYING ANGLES, 84

COROLLARY 3-1, 33–34

COROLLARY 6-4.1, 71

COROLLARY 11-1.1, 110

COROLLARY 12-6.1, 127

COROLLARY 14-1.1, 152–53

COROLLARY 15-1.1, 164

COROLLARY 15-2.1, 169

COROLLARY 17-2.1, 189

COROLLARY 17-2.3, 189, 199

COROLLARY 17-2.4 (THALE'S THEOREM),
    189, 204

CORRESPONDING ANGLES, 93–95, 99,
    101–3, 106

COSINE, 215–20

CYCLIC POLYGONS, 228–29, 242

"DEFINE YOUR TERMS," 13–14

DEFINITIONS, 24

DEGREES, 4, 50–51

DIAGONALS, 5
    CONVEX POLYGONS, 225
    QUADRILATERALS, 119, 124–28
    SQUARES, 5–6, 8, 12
    TRAPEZOIDS, 142

DIAMETER, 39, 51

DISTANCE FROM POINT TO LINE, 79–80

DISTRIBUTIVE LAW, 134

DOUBLE INFINITY, 42

EGYPTIANS, ANCIENT, 2–3, 91, 131

ELEMENTS, THE (EUCLID), 10

ENDPOINTS, 32–33, 71

ENLARGEMENTS, 157–58

EQUIDISTANT, 71, 87, 227

EQUILATERAL TRIANGLES, 112, 153, 231
    DEFINITION, 112
    EXERCISES, 90

EQUIVALENTS, 23–24, 25, 102

EUCLID (EUCLIDIAN GEOMETRY), 10–11, 26,
    27, 45, 77, 81, 96, 148

EXERCISES
    ANGLES, 58, 245
    AREA, 146, 248
    BASIC TERMS, 28, 244
    CIRCLES, 46, 194, 245, 249
    CLASSICAL CONSTRUCTIONS, 90, 246
    GEOMETRIC MEAN, 204, 250
    GOLDEN TRIANGLE, 212, 250
    INEQUALITIES IN TRIANGLES, 82, 246
    INTERSECTION PROBLEM, 98, 247
    LINE NUMBERS, 36, 244
    PARALLEL, 108, 247
    POLYGONS, 242, 251
    PYTHAGOREAN THEOREM, 156, 248
    QUADRILATERALS, 130, 248
    RIGHT TRIANGLES, 222, 250–51
    SCALING, 182, 249
    SIMILARITY, 174, 249
    SOLUTIONS, 244–51
    TRIANGLES, 74, 116, 245, 247

EXTERIOR ANGLES, 76–77, 79, 110
    DEFINITION, 76
    EXERCISES, 82
    THEOREM 7-1, 77
    THEOREM 9-1, 94–95
    THEOREM 21-1, 225

FARMING, DETERMINING AREA, 131–34
FIBONACCI SEQUENCE, 212
FLAT MIRROR, 114–15

GAUSS, CARL FRIEDRICH, 96, 235
GEOMETRICAL STATEMENTS, 19
GEOMETRIC MEAN, 195–204, 206
    EXERCISES, 204, 250
GEOMETRY
    BASIC INGREDIENTS, 13–27
    DEFINITION, 1
    NATURE AND BRIEF HISTORY OF, 1–11
GOLDEN RATIO, 207, 232
GOLDEN RECTANGLES, 211–12, 213
GOLDEN TRIANGLES, 205–12
    EXERCISES, 212, 250
    THEOREM 19-1, 208
    THEOREM 19-2, 209
    THEOREM 19-3, 210
GREEKS, ANCIENT, 7–11, 211

HALF LINES, 31
HALF-PLANES, 53–54
HEPTAGONS, 234–35
HEXAGONS, 223, 234
HYPERBOLOID, 96–97

HYPOTENUSE, 79, 113, 144
    PYTHAGOREAN THEOREM, 147–48,
        150–51, 153–54
HYPOTHESIS, 21

IF-THEN STATEMENTS, 21–22, 24, 28
INCIRCLES, 194
INDIRECT PROOFS, 25, 72
INEQUALITIES IN TRIANGLES, 36, 75–82
    DEFINITION, 76
    DISTANCE FROM POINT TO LINE,
        79–80
    EXERCISES, 82, 246
    EXTERIOR ANGLES, 76–77, 82
    THEOREM 7-1, 77
    THEOREM 7-2, 78
    THEOREM 7-3, 79
    THEOREM 7-4, 80
INSCRIBED ANGLES, 187–89
INTERSECTION PROBLEM, 91–98
    EXERCISES, 98, 247
    THEOREM 9-1, 94–95
    TRANSVERSAL, 93–95
ISOSCELES TRIANGLES, 68–71, 184, 187
    DEFINITION, 68, 70
    EXERCISES, 82, 146, 212
    GOLDEN, 205–12
    HYPOTENUSE OF, 113, 144, 153
    THEOREM 6-3, 69
    THEOREM 6-4, 71
    THEOREM 7-2, 78

KITCHENWARE, USE OF GEOMETRY, 2

LEGS, 68, 135, 144, 150–51, 154

LEMMAS, 162–63

LINES, 17
    DISTANCE FROM POINT TO, 79–80
    EXERCISES, 36, 244
    LACK OF DEFINITION, 15
    NUMBER, 29–31
    THEOREMS, 20–23

LINE SEGMENTS, 32, 44, 119, 165
    DEFINITIONS, 24, 32, 39, 118
    EXERCISES, 90

LOBACHEVSKY, NIKOLAI, 96

MEASUREMENTS, 1–7, 35

MEDIANS, 70

MIDPOINTS, 69–70, 77, 185
    EXERCISES, 182

MIRROR IMAGES, 67–68, 114–15

NATURE AND GEOMETRY, 1–11

NEGATION, 25

NONCOLLINEAR POINTS, 20, 21, 24,
    60, 92
    CIRCLE THROUGH THREE, 87–89, 105

NON-EUCLIDIAN GEOMETRY, 96

NUMBER LINE, 29–31

OBTUSE ANGLES, 52, 58, 220

OCTAGONS, 223

ORDER, 31

PARALLEL, 99–108, 119
    DEFINITION, 101
    EXERCISES, 108, 247
    POSTULATE 10, 99–102

THEOREM 10-1, 103
THEOREM 10-2, 104–5
THEOREM 10-3, 104–5
THEOREM 10-4, 107

PARALLELOGRAMS, 120–28
    AREA, 141–44, 146, 248
    EXERCISES, 130, 248
    THEOREM 12-2, 121
    THEOREM 12-4, 122–23
    THEOREM 12-5, 124
    THEOREM 12-6, 126–27

PARALLEL POSTULATE, 102–6, 109, 111, 121,
    208

PENTAGONS, 223, 232–33

PERPENDICULAR ANGLES, 56, 69–71

PERPENDICULAR BISECTORS, 70–71, 73,
    88–89, 184
    DEFINITION, 70
    EXERCISES, 90
    OF SEGMENT, 85

PERPENDICULARS AND PARALLELS, 104–5

PERPENDICULAR THROUGH A POINT, 86

PHI, 207

PI, 238

PIE SLICES, 4

PLANES, 16, 17
    HALF-, 53–54
    LACK OF DEFINITION, 15
    POSTULATES, 18–19
    THEOREMS, 20–23

POINTS, 17
    CIRCLES, 42
    COLLINEAR, 20, 31–32
    DEFINITION, 15
    DISTANCE FROM POINT TO LINE, 79–80

OF INTERSECTION, 40, 99
LACK OF DEFINITION, 14, 15
MIDPOINTS. SEE MIDPOINTS
NONCOLLINEAR. SEE NONCOLLINEAR
    POINTS
PERPENDICULAR THROUGH A, 86
POSTULATES, 18–19
THEOREMS, 20–23

POLYGONS, 223–42
    CIRCLES AROUND, 228–29
    DEFINITION, 226
    EXERCISES, 242, 251
    FITTING INTO CIRCLES, 230–31
    AS NEAR CIRCLES, 236–40
    POSTULATE 12, 224
    THEOREM 21-1, 225–26
    THEOREM 21-2, 229
    THEOREM 21-3, 230

POSTULATES, 18–19, 26
    EXERCISES, 28, 244

POSTULATE 1, 18

POSTULATE 2, 18

POSTULATE 3, 18

POSTULATE 4, 18

POSTULATE 5 (RULER POSTULATE), 30

POSTULATE 6, 43–44

POSTULATE 7 (PROTRACTOR POSTULATE),
    53–54, 66, 208

POSTULATE 8 (SAS), 65–66, 73, 171

POSTULATE 9 (ASA), 65–66, 73

POSTULATE 10, 99–102

POSTULATE 11, 134

POSTULATE 12, 224

PRIMITIVE, 202

PROPORTIONS, 159–64

PROTRACTOR POSTULATE (POSTULATE 7),
    53–54, 66, 208

PROTRACTORS, 51–54
    MEASURE AN ANGLE, 52–53

PYTHAGORAS, 7–8, 27

PYTHAGOREAN THEOREM, 147–56
    EXERCISES, 156, 248
    PROOF #1, 149–51
    PROOF #2, 170–71
    PROOF #3, 179–80
    PROOF #4, 200
    IN REAL LIFE, 152

PYTHAGOREAN TRIPLES, 201–3

QUADRATIC FORMULA, 206–7

QUADRILATERALS, 118–20
    AREA, 141–42
    DEFINITION, 118
    EXERCISES, 130, 248
    GALLERY OF, 128
    THEOREM 12-3, 121
    THEOREM 12-4, 122–23
    THEOREM 12-6, 126–27

RADIUS, 39–41, 43–44, 184, 238
    EXERCISES, 46

RATIOS, 159–64, 214–15
    ASPECT, 160, 162, 182
    GOLDEN, 207, 232

RAYS, 31–32, 48, 49, 52–54

RECTANGLES, 3, 12, 127
    AREA, 132–34, 146
    GOLDEN, 211–12, 213
    PARALLELOGRAMS, 120
    PROPORTIONS, 160–62
    SCALING, 158, 160, 175, 182

REFLEX ANGLES, 240

RHOMBUS, 127, 128

RIEMANN, BERNHARD, 96

RIGHT ANGLES, 56, 79, 110, 112–13

RIGHT TRIANGLES, 112–13, 213–22
    30°-60°, 112–13, 153–54
    45°-45°, 144, 153
    AREA, 135–38
    CONGRUENCE TEST, 113
    DEFINITION, 112
    EXERCISES, 222, 250–51
    SINE AND COSINE, 215–20
    THEOREM 11-4, 112–13
    THEOREM 11-5, 113
    THEOREM 15-3, 169
    TRIGONOMETRIC, 221

SCALING, 175–82
    EXAMPLE OF SCALE MODEL, 181
    EXERCISES, 182, 249
    RECTANGLES, 158, 160, 175
    THEOREM 16-1, 176
    THEOREM 16-2, 177–80
    TRIANGLES, 175–77

SEGMENTS. SEE ALSO LINE SEGMENTS
    PERPENDICULAR BISECTORS OF, 85

SIDE-ANGLE-SIDE (POSTULATE 8), 65–66,
    73, 171

SIDE-SIDE-SIDE (THEOREM 6-5),
    72–73, 127

SIMILARITY, 157–74
    CONSTRUCTION, 165–69
    DEFINITION, 166
    EXERCISES, 174, 249
    PROPORTIONS, 159–64
    THEOREM 15-1, 163

THEOREM 15-2, 167–68
THEOREM 15-3, 169, 170
THEOREM 15-4, 171
THEOREM 15-5 (SIMILARITY SSS),
    172–73

SIMILARITY SAS (THEOREM 15-4), 171, 173

SIMILARITY SSS (THEOREM 15-5), 172–73

SINE, 215–20, 238

SPACE, 16
    LACK OF DEFINITION, 14, 15
    POSTULATES, 18–19

SQUARES, 127–29
    DIAGONALS, 5–6, 8, 12
    FITTING POLYGONS INTO, 231
    SCALING, 175, 182
    TILTED, 3, 12

SQUARE DEALING, 129

SQUARE ROOT OF 2, 6–8

STATEMENTS
    CONTRAPOSITIVES, 25, 28, 95, 104
    CONVERSE, 22–23, 28
    EXERCISES, 28
    GEOMETRICAL, 19

STRAIGHTEDGE, 9, 83, 235
    EXERCISES, 90

SUPPLEMENTARY ANGLES, 57, 106

TANGENTS, 190–93
    EXERCISES, 194, 249

THALE'S THEOREM, 189, 204

THEOREMS, 20–23

THEOREM 2-1, 20–21

THEOREM 3-1, 33

THEOREM 3-2, 34

THEOREM 4-1, 40–41

THEOREM 4-2, 42

THEOREM 5-1, 54–55

THEOREM 5-2, 55

THEOREM 5-3, 56

THEOREM 6-1, 66–67

THEOREM 6-2, 67

THEOREM 6-3, 69

THEOREM 6-4, 71

THEOREM 6-5 (SSS), 72–73, 127

THEOREM 7-1, 77

THEOREM 7-2, 78, 82

THEOREM 7-3, 79

THEOREM 7-4 (TRIANGLE INEQUALITY),
      80, 82

THEOREM 8-0, 105

THEOREM 8-1, 88–89

THEOREM 9-1, 94–95, 100

THEOREM 10-1, 103

THEOREM 10-2, 104–5

THEOREM 10-3, 104–5

THEOREM 10-4, 107, 111

THEOREM 11-1, 110

THEOREM 11-2, 110

THEOREM 11-3 (AAS), 111

THEOREM 11-4, 112–13

THEOREM 11-5, 113

THEOREM 12-1, 110

THEOREM 12-2, 121

THEOREM 12-3, 121

THEOREM 12-4, 122–23

THEOREM 12-5, 124

THEOREM 12-6, 126–27

THEOREM 13-1, 135

THEOREM 13-2, 136–37

THEOREM 13-3, 139–40

THEOREM 13-4, 142

THEOREM 14-1, 148

THEOREM 15-1, 163

THEOREM 15-2, 167–68

THEOREM 15-3, 169, 170

THEOREM 15-4 (SIMILARITY SAS), 171, 173

THEOREM 15-5 (SIMILARITY SSS), 172–73

THEOREM 16-1, 176

THEOREM 16-2, 177–80

THEOREM 17-1, 185

THEOREM 17-2, 188

THEOREM 17-3, 190–91, 193

THEOREM 17-4, 191

THEOREM 19-1, 208

THEOREM 19-2, 209

THEOREM 19-3, 210

THEOREM 20-1 (LAW OF SINES), 218

THEOREM 20-2 (LAW OF COSINES), 219

THEOREM 21-1, 225–26

THEOREM 21-2, 229

THEOREM 21-3, 230

"TILT," 93, 95, 97

TILTED SQUARES, 3, 12

TOWN SQUARES, 120

TRANSVERSALS, 93–95, 99–102, 106

TRAPEZOIDS, 120, 128
    AREA, 141–44
    EXERCISES, 156, 248

TRIANGLES, 59–74, 109–16, 194
    ACUTE, 79, 214, 218–19
    ALTITUDE, 136–40, 153, 169,
        176–77, 218
    ANGLES, 23
    APEX, 68–70, 205
    AREA, 135–40
    BASE OF, 68–70
    COMPARING, 60–61
    CONGRUENCE, 61–66
    DEFINITIONS, 24, 60, 70–71
    EQUILATERAL. SEE EQUILATERAL
        TRIANGLES
    EXAMPLE OF FLAT MIRROR, 114–15
    EXERCISES, 74, 116, 245, 247
    FITTING POLYGONS INTO, 231
    GOLDEN. SEE GOLDEN TRIANGLES
    INEQUALITIES IN. SEE INEQUALITIES IN
        TRIANGLES
    ISOSCELES. SEE ISOSCELES TRIANGLES
    PROPORTIONS, 162–64
    PYTHAGOREAN THEOREM, 152–53
    RIGHT. SEE RIGHT TRIANGLES
    SCALING, 175–77, 182
    THEOREM 6-1, 66–67
    THEOREM 6-2, 67

    THEOREM 6-3, 69
    THEOREM 6-4, 71
    THEOREM 6-5, 72–73
    THEOREM 11-1, 110
    THEOREM 11-2, 110
    THEOREM 11-3, 111
    THEOREM 11-4, 112–13
    THEOREM 11-5, 113
    THEOREM 13-2, 136–37
    THEOREM 13-3, 139–40

TRIGONOMETRIC, 221

TRI-SQUARES, 120

T-SQUARES, 120

TWO-ANGLE TEST, 169

UNDEFINED TERMS, 14

UNIVERSAL RULER, 29–30, 35

UNSTATED ASSUMPTIONS, 26

VERIFIABLE STATEMENTS, 19

VERTEX ANGLE, 49, 52–53, 55, 60,
    63, 118–20, 162

VERTICAL ANGLES, 55–56, 82, 94, 106

WHOLE NUMBERS, 7

Discover the World with Cartoons!

Refreshingly humorous and thorough guides from the author of the bestselling Cartoon Guide series.

THE CARTOON GUIDE TO BIOLOGY

Uses simple, clear, humorous illustrations to make biology's most complex concepts understandable and entertaining.

THE CARTOON GUIDE TO ALGEBRA

Covers algebra essentials, including linear equations, polynomials, quadratic equations, and graphing techniques. Offers a concise overview of algebra's history and its many practical applications in modern life.

THE CARTOON GUIDE TO CALCULUS

Entertainingly teaches all of the course essentials, including functions, limits, and derivatives, as well as more advanced integrals, series, and convergence tests. Concludes with a quick but thorough look at the theoretical basis and history of calculus.

THE CARTOON HISTORY OF THE MODERN WORLD PART 1

Beginning with Columbus's arrival on America's shores, traces the story of humanity through to the United States' emergence as a world superpower, the technological revolution, and the resurgence of militant religion.

THE CARTOON HISTORY OF THE MODERN WORLD PART 2

From the beginning of the Enlightenment to the environmental crisis of today's world. A unique spin on the history, personalities, and big topics that have shaped our world over the past five centuries.

THE CARTOON GUIDE TO CHEMISTRY

A course in college-level chemistry, the book covers electrochemistry, organic chemistry, biochemistry, environmental chemistry, physics as chemistry, and much more.

THE CARTOON GUIDE TO THE ENVIRONMENT

Covers the main topics of environmental science: chemical cycles, life communities, food webs, agriculture, human population growth, sources of energy and raw materials, waste disposal and recycling, cities, pollution, deforestation, ozone depletion, and global warming.

THE CARTOON GUIDE TO STATISTICS

Covers all the central ideas of modern statistics: probability in gambling and medicine, random variables, Bernoulli trials, hypothesis testing, confidence interval estimation—all explained in simple, clear, and, yes, funny illustrations.

THE CARTOON GUIDE TO GENETICS

Ease your way through Mendelian genetics, molecular biology, and the basics of genetic engineering.

THE CARTOON HISTORY OF THE UNITED STATES

From the first English colonies to the Gulf War and the S&L debacle, Larry Gonick spells it all out from his unique cartoon perspective.

THE CARTOON GUIDE TO SEX

Frank, informative, and written with Larry Gonick's characteristic comic verve and scientific accuracy, this book gives a comprehensive discussion of the spectrum of human sexuality.

THE CARTOON GUIDE TO PHYSICS

Explains many complex ideas through simple, clear, and funny illustrations: velocity, acceleration, explosions, electricity and magnetism, circuits—even a taste of relativity theory—and much more.